Advances in Spatial Science

Springer-Verlag Berlin Heidelberg GmbH

Titles in the Series

C.S. Bertuglia, M.M. Fischer and *G. Preto* (Eds.)
Technological Change, Economic Development and Space
XVI, 354 pages. 1995. ISBN 3-540-59288-1 (out of print)

H. Coccossis and *P. Nijkamp* (Eds.)
Overcoming Isolation
VIII, 272 pages. 1995. ISBN 3-540-59423-X

L. Anselin and *R.J.G.M. Florax* (Eds.)
New Directions in Spatial Econometrics
XIX, 420 pages. 1995. ISBN 3-540-60020-5 (out of print)

H. Eskelinen and *F. Snickars* (Eds.)
Competitive European Peripheries
VIII, 271 pages. 1995. ISBN 3-540-60211-9

J.C.J.M. van den Bergh, P. Nijkamp and *P. Rietveld* (Eds.)
Recent Advances in Spatial Equilibrium Modelling
VIII, 392 pages. 1996. ISBN 3-540-60708-0

P. Nijkamp, G. Pepping and *D. Banister*
Telematics and Transport Behaviour
XII, 227 pages. 1996. ISBN 3-540-60919-9

D.F. Batten and *C. Karlsson* (Eds.)
Infrastructure and the Complexity of Economic Development
VIII, 298 pages. 1996. ISBN 3-540-61333-1

T. Puu
Mathematical Location and Land Use Theory
IX, 294 pages. 1997. ISBN 3-540-61819-8

Y. Leung
Intelligent Spatial Decision Support Systems
XV, 470 pages. 1997. ISBN 3-540-62518-6

C.S. Bertuglia, S. Lombardo and *P. Nijkamp* (Eds.)
Innovative Behaviour in Space and Time
X, 437 pages. 1997. ISBN 3-540-62542-9

A. Nagurney and *S. Siokos*
Financial Networks
XVI, 492 pages. 1997. ISBN 3-540-63116-X

M.M. Fischer and *A. Getis* (Eds.)
Recent Developments in Spatial Analysis
X, 434 pages. 1997. ISBN 3-540-63180-1

R.H.M. Emmerink
Information and Pricing in Road Transportation
XVI, 294 pages. 1998. ISBN 3-540-64088-6

P. Rietveld and *F. Bruinsma*
Is Transport Infrastructure Effective?
XIV, 384 pages. 1998. ISBN 3-540-64542-X

P. McCann
The Economics of Industrial Location
XII, 228 pages. 1998. ISBN 3-540-64586-1

L. Lundqvist, L.-G. Mattsson and *T.J. Kim* (Eds.)
Network Infrastructure and the Urban Environment
IX, 414 pages. 1998. ISBN 3-540-64585-3

R. Capello, P. Nijkamp and *G. Pepping*
Sustainable Cities and Energy Policies
XI, 282 pages. 1999. ISBN 3-540-64805-4

M.M. Fischer and *P. Nijkamp* (Eds.)
Spatial Dynamics of European Integration
XII, 367 pages. 1999. ISBN 3-540-65817-3

M.M. Fischer, L. Suarez-Villa and *M. Steiner* (Eds.)
Innovation, Networks and Localities
XI, 336 pages. 1999. ISBN 3-540-65853-X

J. Stillwell, S. Geertman and *S. Openshaw* (Eds.)
Geographical Information and Planning
X, 454 pages. 1999. ISBN 3-540-65902-1

G.J.D. Hewings, M. Sonis, M. Madden and *Y. Kimura* (Eds.)
Understanding and Interpreting Economic Structure
X, 365 pages. 1999. ISBN 3-540-66045-3

A. Reggiani (Ed.)
Spatial Economic Science
XII, 457 pages. 2000. ISBN 3-540-67493-4

D.G. Janelle and *D.C. Hodge* (Eds.)
Information, Place, and Cyberspace
XII, 381 pages. 2000. ISBN 3-540-67492-6

P.W.J. Batey and *P. Friedrich* (Eds.)
Regional Competition
VIII, 290 pages. 2000. ISBN 3-540-67548-5

B. Johansson, Ch. Karlsson and *R.R. Stough* (Eds.)
Theories of Endogenous Regional Growth
IX, 428 pages. 2001. ISBN 3-540-67988-X

G. Clarke and *M. Madden* (Eds.)
Regional Science in Business
VIII, 363 pages. 2001. ISBN 3-540-41780-X

M.M. Fischer and *Y. Leung* (Eds.)
GeoComputational Modelling
XII, 279 pages. 2001. ISBN 3-540-41968-3

M.M. Fischer and *J. Fröhlich* (Eds.)
Knowledge, Complexity and Innovation Systems
XII, 477 pages. 2001. ISBN 3-540-41969-1

Manfred M. Fischer · Javier Revilla Diez
Folke Snickars

In Association with Attila Varga

Metropolitan Innovation Systems

Theory and Evidence from Three Metropolitan Regions in Europe

With 40 Figures
and 159 Tables

Springer

Professor Dr. Manfred M. Fischer
Vienna University of Economics and Business Administration
Department of Economic Geography & Geoinformatics
Rossauer Lände 23/1
1090 Vienna
Austria

Dr. Javier Revilla Diez
University of Hannover
Department of Economic Geography
Schneiderberg 50
30167 Hannover
Germany

Professor Dr. Folke Snickars
Royal Institute of Technology
Department of Regional Planning
10044 Stockholm
Sweden

ISBN 978-3-642-07548-3

Library of Congress Cataloging-in-Publication Data applied for
Die Deutsche Bibliothek – CIP-Einheitsaufnahme
Fischer, Manfred M.: Metropolitan Innovation Systems: Theory and Evidence from Three Metropolitan Regions in Europe / Manfred M. Fischer; Javier Revilla Diez; Folke Snickars. In Assoc. with Attila Varga. – Berlin; Heidelberg; New York; Barcelona; Hong Kong; London; Milan; Paris; Singapore; Tokyo: Springer, 2001
(Advances in Spatial Science)
ISBN 978-3-642-07548-3 ISBN 978-3-662-04630-2 (eBook)
DOI 10.1007/978-3-662-04630-2

http://www.springer.de

Originally published by Springer-Verlag Berlin Heidelberg New York in 2001
Softcover reprint of the hardcover 1st edition 2001

Hardcover-Design: Erich Kirchner, Heidelberg

SPIN 10835130 42/2202-5 4 3 2 1 0 – Printed on acid-free paper

Preface

This book presents the findings of a comparative study of three European metropolitan regions: Vienna, Barcelona and Stockholm. The heart of the work consists of empirical studies carefully designed and developed in order to identify the main actors and mechanisms supporting technological innovation in each of the metropolitan regions. The authors have also highlighted the similarities and differences across regions and countries, investigating how these came to be, and discussing the possible implications.

The introductory as well as the concluding Chapter was written by Manfred M. Fischer who, assisted by Attila Varga, was also responsible for Chapter 2 on the Metropolitan Region of Vienna. Javier Revilla Diez contributed Chapter 3 on the Barcelona Metropolitan Region. Folke Snickars has provided Chapter 4 which examines the Metropolitan Region of Stockholm and. All authors have reviewed and commented on the whole contents so that the volume represents a collective endeavour which has been rendered as homogeneous as possible. A particular effort has been made to ensure that the study is based on a common conceptual framework.

The project that led to this book was an integral part of the second stage of a larger research programme concerning 'Technological Change and Regional Development in Europe' sponsored by the German Research Foundation and ably managed by Ludwig Schätzl (University of Hanover), the overall project co-ordinator. The research has been conducted through a partnership of five research institutions: the University of Hanover, the Polytechnic University of Catalunia, the Royal Institute of Technology Stockholm, the Vienna University of Economics and Business Administration, and the Austrian Academy of Sciences.

The authors of this volume wish to thank Vera Mayer, Walter Rohn (both of the Austrian Academy of Sciences), Ingo Liefner (University of Hanover) and Olof Seidel (The Royal Institute of Technology Stockholm) for providing fundamental help in conducting the postal surveys. The design of the surveys has drawn heavily on the questionnaires developed by Ludwig Schätzl (University of Hanover), Rolf Sternberg (University of Cologne), Max Fritsch (Technical University of Freiberg), Frieder Meyer-Krahmer and Knut Koschatzky (Frauenhofer Institute for Systems and Innovation Research) in the first stage of the above research programme of the German Research Foundation with a focus on Central European regions. The use of professional translation in the production of the questionnaire forms and the exercise of central control (University of Hanover) over the sampling methodologies and production of the questionnaires has been crucial to the optimisation of this European metropolitan innovation survey.

We wish to acknowledge the support provided by the German Research Foundation, the Institute for Urban and Regional Research at the Austrian Academy of Sciences, the Department of Economic Geography and Geoinformatics at Vienna University of Economics and Business Administration, the Institute of Geography at the University of Hanover and the Department of Infrastructure and Planning at the Royal Institute of Technology Stockholm. We would like to thank Thomas Seyffertitz for his capable assistance in co-ordinating the various stages of preparation of the book. Finally, we would like to thank Angela Spence for providing expert editorial assistance. Her care and attention to the linguistic editing and indexing, together with her efforts in the production of the final copy have considerably enhanced the quality of the work presented here.

March, 2001

Manfred M. Fischer
Vienna University of Economics and Business Administration

Javier Revilla Diez
University of Hanover

Folke Snickars
Royal Institute of Technology Stockholm

Contents

1 Systems of Innovation: An Attractive Conceptual Framework for Comparative Innovation Research

Today, it is widely recognised that technological change is the primary engine of economic development. Innovation – the heart of technological change – is a process that depends essentially upon the accumulation and development of a wide variety of relevant knowledge. Certainly individual firms play a crucial role in the development of specific innovation, but the process that nurtures and disseminates technological change in the economy involves a complex web of interactions among a range of firms, organisations and institutions.

The innovation systems approach has recently received considerable attention and is being adopted as conceptual framework for the comparative study reported here. The approach contrasts with those previously adopted, such as the OECD's traditional approach to technological change and innovation, which focuses on the R&D system in a narrower sense, primarily analysing resource inputs and outputs. An over-narrow focus on R&D tends to overlook the importance of other types of innovative effort in the business sectors and, thus, the innovative performance of low-tech sectors of the economy.

The main objective of this introductory chapter is to provide greater understanding of the systems of innovation approach, introducing the reader to some of the central concepts of this useful and flexible framework for innovation analysis. Section 1.1 stresses the importance of adopting a broad view of innovation in order to emphasise its interactive nature. As discussed in Section 1.2, knowledge creation and dissemination are at the very centre of focus of the systems approach. These concepts recur in later discussion of the nature of the conceptual framework.

A system of innovation may be thought of as a set of actors, such as firms, organisations and institutions, that interact in the generation, diffusion and use of new – and economically useful – knowledge in the production process. Section 1.3 addresses conceptual issues from which innovation studies might benefit. Particular stress is laid on showing how the conceptual core of the system approach needs to be specified in broad terms in order to provide a useful framework for the systematic and rigorous comparison of innovation systems.

Localised systems, such as metropolitan systems, build on some kind of spatial proximity. However, current research on systems of innovation focuses almost exclusively on the national scale (see, for example, Lundvall 1992; Nelson 1993; OECD 1994; Edquist 1997a). As argued in Section 1.4, there is *no a priori reason* to privilege this particular spatial scale, irrespective of time and place. A strong case is made here for the importance of the regional scale as an appropriate mode for analysis. In Europe it seems to be regional rather than national systems of innovation that matter. Localised input-output relations between the actors of the system, knowledge spillovers and their untraded interdependencies lie at the heart of the arguments. There is increasing evidence that metropolitan regions tend to be the principal engines of industrial innovation and growth in national economies. But there is no comparative evidence across metropolitan regions based on the use of a common and more rigid methodology (see Section 1.5). The present book makes an attempt to fill this gap. The research design of the study is briefly described in Section 1.6, while Section 1.7 provides a brief guide to this volume.

1.1 Innovation as an Interactive Process

Technological change is a complex process whose workings are not yet fully understood. This complexity stems partially from the diverse set of phenomena that are subsumed under the term innovation. Bienaymé (1986), for example, distinguishes between a) product innovations; b) innovations destined to resolve, circumvent or eliminate a technical difficulty in manufacture or to improve services; c) innovations for the purpose of saving inputs (e.g. energy conservation, automation), and d) innovations to improve the working conditions. These very different phenomena have made generalisation difficult (Malecki 1997). For a long time, thinking about technological change and innovation was determined by linear models - in the 1950s and 1960s by the technology-push and then the need-pull model. In the former, the development, production and marketing of new technology - defined by Mansfield et al. (1982) as consisting of a pool or set of knowledge - was assumed to follow a well defined time sequence which began with basic and applied research activities, involved a product development stage, and then led to production and possibly commercialisation. In the second model, this linear sequential process emphasised demand and markets as the source of ideas for R&D activities. These models have guided the formulation of national R&D policies in the past, but have come under increasing attack in recent years for several reasons, not least due to the absence of feedback loops between the downstream (market-related) and upstream (technology-related) phases of innovation. The current intensification of competition and shorter product life cycles are requiring a closer integration of R&D with other phases of the innovation process.

This criticism has led to a broader view of the innovation process, stressing its interactive nature. The emerging innovation theory emphasises the central role of feedback effects between the downstream and upstream phases of innovation, as well as the numerous interactions between science, technology and innovation related activities within and among firms. Through interactions and feedbacks different pieces of knowledge become combined in new ways and, in some cases, new knowledge is created.

Fig. 1.1 represents what is referred to as the 'chain-linked model' (Kline and Rosenberg 1986; OECD 1992; Malecki 1997). The innovation process at the firm level is portrayed as a set of activities linked to one another through complex feedback loops. The process can be visualised as a chain, starting with the perception of a new market opportunity and/or a new invention based on novel pieces of scientific and/or technological knowledge followed by the analytical design for a new product or process and testing, redesign and production, and distribution and marketing. Short feedback loops link each downstream phase in the central chain with the phase immediately preceding it. Longer feedback loops link perceived market demand and product users with phases upstream. The second set of relationships visualised in Fig.1.1 link the innovation process embedded in the firm with its firm-specific knowledge base, the general scientific and technological knowledge pool and with research activities.

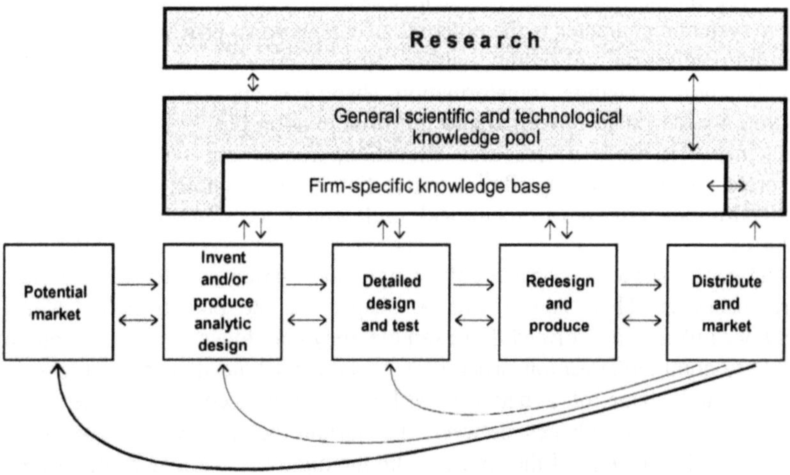

Source: Adapted with minor changes from Kline and Rosenberg (1986), Myers and Rosenbloom (1996), Malecki (1997).

Fig. 1.1 An interactive model of the innovation process: Feedbacks and interactions (Fischer 1999)

The model combines two types of interaction. The first concerns processes that occur through new forms of product development within the firm and create appropriate feedback relationships (see, for example, Nonaka and Takeuchi 1995). The second refers to relationships external to a given firm, for example with customers, suppliers of inputs (including finance and knowledge), research institutions and even competitors. Co-operation can take place with various mixes of internal and external actors. Under this model, technological innovation is seen as the result of a complex interplay among various actors with partly common and partly conflicting interests. Technological progress is thus dependent on how they interact with each other, internally and externally.

In recent years, new forms of inter-firm agreement bearing on technology have developed alongside the traditional means of technology transfer – licensing and trade in patents – and often have become the most important way for firms, regions and nation-states to gain access to new knowledge and key technologies. The network form of governance can overcome market imperfections as well as the rigidities of vertically integrated hierarchies. The limitations of these two modes of transaction in the context of knowledge and innovation diffusion have pushed interfirm agreements to the forefront of corporate strategy in the last few decades (Chesnais 1988).

There are many definitions of *innovation networks* (see De Bresson and Amesse 1991; Freeman 1991). However, the one offered by Tijssen captures the most important features of the network mode. He suggests defining a network as "an evolving mutual dependency system based on resource relationships in which their systemic character is the outcome of interactions, processes, procedures and institutionalisation. Activities within such a network involve the creation, combination, exchange, transformation, absorption and exploitation of resources within a wide range of formal and informal relationships" (Tijssen 1998, p.792). In a network mode of resource allocation, transactions occur neither through discrete exchanges nor by administrative fiat, but through networks of individuals or institutions, engaged in reciprocal, preferential and supportive actions (Powell 1990).

Networks show a considerable range and variety, differing according to circumstances. Their nature will be shaped by the objectives for which the network linkages are formed. For example, they may focus on a single point of the R&D-to-commercialisation process or cover the whole innovation process. The content and shape of a network will also differ according to the nature of relationships and linkages between the various actors involved (see Chesnais 1988). At the one end of the spectrum lie the highly formalised relationships. The formal structure may consist of regulations, contracts and rules that link actors and activities with varying degrees of constraint. At the other end are the network relations of a mainly informal nature, linking actors through open chains. Such relations are very hard to measure (Freeman 1991). When inter-firm transactions are small in scale, variable and unpredictable, requiring face-to-face contact, then network formation will tend to focus on the close geographical proximity of the partners involved (Storper 1997).

For firms, networks represent a response to quite specific circumstances. Where complementarity is a prerequisite for successful innovation, network agreements may be formed in response to specific proprietary tacit knowledge. The exchange of such complementary assets can take place only through very close contacts and personalised, generally localised, relationships. When technology is evolving rapidly, another reason for preferring a network mode is that it provides a far higher degree of flexibility, reversibility and risk-sharing (OECD 1992). Inter-firm agreements are easier to dissolve than internal developments or mergers. Porter and Fuller (1986) stress speed as being among the advantages of networks over acquisition or internal development through arm's length relationships. This advantage is becoming increasingly important as product life cycles have shortened and competition has intensified. High R&D costs may be another distinct reason for networking and can force management, especially in smaller firms, to pool resources with other firms, in some cases even with competitors (OECD 1992).

1.2 Knowledge Creation and Diffusion

Recognition of the interactive nature of the innovation process has resulted in the breakdown of the earlier distinction between innovation and diffusion. The creation of knowledge and its assimilation are seen as part of a single process and are at the heart of the conceptual framework adopted in the present analaysis. Firms need to absorb, create and exchange knowledge interdependently. In other words, innovation and diffusion usually emerge as a result of an interactive and collective process within a web of personal and institutional connections which evolve over time.

Knowledge transfer may occur through disembodied or equipment-embodied diffusion. The latter is the process by which innovations spread in the economy through the purchase of technology-intensive machinery, such as computer-assisted systems, components and other equipment. Disembodied technology diffusion refers to the process where technology and knowledge spread through other channels not embodied in machinery (OECD 1992). This type of knowledge transfer may occur via descriptions of new products or production processes found in catalogues, publications or patent applications, but also via seminars and conferences, and R&D personnel turnover. It can also be the by-product of mergers and acquisitions, joint ventures or other forms of inter-firm co-operation.

Two notions are central to an understanding of disembodied technology diffusion: the first is that of absorption capacity and the second that of knowledge spillovers. The *absorption capacity* of firms and other organisations refers to the ability to learn, assimilate and use knowledge developed elsewhere through a process that involves substantial investments, especially of an intangible nature (Cohen and Levinthal 1989). This capacity depends crucially on the learning experience, which in turn may be enhanced by in-house R&D activities. The concept of absorption capacity implies that in order to have access to a piece of knowledge developed elsewhere, it is necessary to have undertaken R&D on something similar (Saviotti 1998). Thus, R&D may be viewed as serving a dual, but strongly interrelated role: firstly, developing new products and production processes, and secondly, enhancing the learning capacity.

Firms, especially smaller firms, that lack appropriate in-house R&D facilities have to develop and enhance their absorption capacity by other means, such as learning from customers and suppliers, interacting with other firms and taking advantage of knowledge spillovers from other firms and organisations (Lundvall 1988). These sources provide the know-why (i.e. procedural knowledge), know-how (i.e. skills and competences) and know-what (i.e. factual knowledge) important for entrepreneurial success (Johannisson 1991; Malecki 1997). Network arrangements of different kinds provide a firm with the assistance necessary to take advantage of outside knowledge.

Knowledge spillovers (i.e. knowledge created by one firm which can be used by another without compensation or with compensation less than the value of knowledge) arise because knowledge and innovation are only partially excludable and non-competing goods (Romer 1990). Lack of *excludability* implies that knowledge producers have difficulty in fully appropriating the returns or benefits and preventing other firms from utilising the knowledge without compensation (Teece 1986). Patents and other devices, such as lead times and secrecy, are a way for knowledge producers to partially capture the benefits related to knowledge creation. It is important to recognise that even a completely codified piece of knowledge cannot be utilised by everyone at zero cost. Only those economic agents who know the code are able to do so (Saviotti 1998).

Through non-rivalry, knowledge distinguishes itself from all other inputs in the production process. *Non-rivalry* means essentially that a new piece of knowledge can be utilised many times and in many different circumstances, for example by combining with knowledge coming from another domain. The interest of the knowledge users is thus best served if innovations, once produced, are made widely available and diffused at the lowest possible cost. This implies an environment rich in knowledge spillovers (OECD 1992).

New understanding of the nature of knowledge associated with technological innovation processes is at the heart of recent conceptual advance. Innovation – in the form of advancing technology – combines two types of knowledge: *codified* (also termed explicit) knowledge drawn from previous experience and *uncodified* (implicit) knowledge which is industry-specific, firm-specific or even individual-specific, and has some degree of tacitness. In each technology there are elements of both tacit and specific knowledge. Following Polanyi (1966), *tacitness* refers to those elements of knowledge which are ill-defined, uncodified and which even those who possess it cannot fully articulate. It differs from person to person, but may to some degree be shared by collaborators who have common experience. Shared knowledge is seldom completely tacit or completely codified (i.e. explicit). In most cases a piece of knowledge can be located between these two extremes. Knowledge is always at least partly tacit in the minds of those who create it. Codification is required because knowledge creation is a collective process that requires complex mechanisms of communication and transfer (Saviotti 1988). With an increase in the tacit components in the firm's knowledge base – due to common practice based on value systems, modes of interpretations and perceptions – knowledge accumulation becomes more experience based, i.e. based on firm specific skills and competences like reliability and reputation. Such forms of knowledge can only be shared, communicated or transferred through network types of relationship.

In an economic system where innovation is crucial for competitiveness, the ability to create knowledge becomes the foundation of innovating firms. Nonaka and Takeuchi (1995) have recently proposed a simple, but elegant model to account for the generation of knowledge in the firm. What they call the 'knowledge-creating company' is based on the organisational interaction between codified (explicit) knowledge and implicit knowledge at the source of innovation. Knowledge creation in an organisation reflects the importance of institutional learning processes and involves two forms of interaction: between tacit and explicit knowledge, and between individuals and the organisation. The interaction between the two forms of knowledge is the key dynamic of knowledge creation in the business organisation. It will bring about four major processes of knowledge conversion that require special learning processes and together constitute knowledge creation (see Fig. 1.2):

- *from tacit to explicit knowledge*, the so-called 'externalisation mode' that holds the key to knowledge creation because it generates new explicit concepts from tacit knowledge. Codification is at the heart of this mode;
- *from explicit to tacit knowledge*, the so-called 'internalisation mode' that is closely related to learning-by-doing and leads to operational/procedural knowledge;
- *from tacit into tacit knowledge*, the so-called 'socialisation mode' that is a process of sharing experiences and thereby creating some sort of novel tacit knowledge, such as technical skills;
- *from explicit to explicit knowledge*, the so-called 'combination mode' that is a process which involves combining different bodies of explicit knowledge in order to create systemic knowledge; a mode that is widely occurring in instructing, training and supervision of the employees.

It is important to note that knowledge is produced by individuals, not by the organisation itself. If the knowledge cannot be shared with others or is not amplified at the group level, it does not move up to the organisational level.

Nonaka and Takeuchi (1995) argue that the core of the creation process of organisational knowledge takes place at the group level, but it is the organisation which provides the enabling conditions. The organisational context is made up of conventions, managerial ideologies, customs, habits and established business practices that facilitate the creation and accumulation of knowledge at the organisational level. Organisational knowledge creation is thus a complex non-linear interactive process characterised by a continuous and dynamic interaction between tacit and explicit forms of knowledge that is shaped by shifts between the above four different modes of knowledge transformation. This knowledge creation process requires the full participation of the workers so that they do not keep their tacit knowledge solely for their own benefit. It also requires stability of the labour force in a firm because only then it is rational for the individual to transfer his/her knowledge to the organisation, and for the firm to diffuse explicit knowledge (Castells 1996). On-line communication along with artificial agents and expert systems have become powerful tools in recent times in helping to manage the complexity of necessary organisational links in the knowledge creation process.

Tacit Knowledge *to* Explicit Knowledge

Tacit Knowledge	Sympathised Knowledge *Socialisation*	Conceptual Knowledge *Externalisation*
from		
Explicit Knowledge	Procedural Knowledge *Internalisation*	Systemic Knowledge *Combination*

Fig. 1.2 Four major processes of knowledge conversion (Nonaka and Takeuchi 1995)

1.3 The Innovation Systems Approach

The innovation systems approach is not a formal theory, but a conceptual framework - a framework in its early stage of development. The idea that lies at the centre of this framework is – as already mentioned above – that the economic performance of territories (regions or countries) depends not only on how business corporations perform, but also on how they interact with each other and with the public sector in knowledge creation and dissemination. Innovating firms operate within a common institutional set-up, and they jointly depend on, contribute to and use a common knowledge infrastructure. Consequently, as discussed in the previous section, the approach places innovation, knowledge creation and diffusion at its very centre. Innovation and knowledge creation are viewed as interactive and cumulative processes contingent on the institutional set-up. It departs from the network school of research (Håkansson 1987) with its emphasis on the institutional set-up, i.e. the role that institutions play in the innovation process (see Edquist and Johnson 1997). The concept of institutions refers at an abstract level to the recurrent patterns of behaviour, socially inherited habits, conventions including regulation, values and routines (Morgan 1997), that assist in regulating life.

A *system of innovation* can be considered to consist of a set of actors or entities such as firms, other organisations and institutions that interact in the generation, use and diffusion of new – and economically useful – knowledge in the production process. At the current stage of development, there is no general agreement on which elements and relations are essential to the conceptual core of the framework and what their precise content is (Edquist 1997b). This leaves room for a conceptual discussion.

Systems that attempt to encompass the whole innovation process may be expected to include four key building blocks that comprise groups of actors sharing some common characteristics and institutions governing the relations within and between the groups (see Fig. 1.3):

- *The Manufacturing Sector*
  This sector is made up of manufacturing firms (the central actors in the system of innovation) and their R&D laboratories that play a fundamental role in performing research and technological development.
- *The Science & Research Sector*
  The science & research sector plays a very important role in technological innovation. It consists of two components: a training component that includes educational and training organisations which act as a source of scientists, engineers, technicians and other skilled workers possessing appropriate skill profiles, and a research component including universities and other research organisations that generate and diffuse knowledge and produce documents in the form of scientific publications. This sector involves various agents (government, private non-profit universities, higher education) that fund and carry out research or offer education.
- *The Innovation Support Units (producer service providers)*
  This sector includes organisations or units within larger organisations which provide assistance or support to industrial firms for the development and/or introduction of new products or processes. This may take any of the following forms: financial, technical advice or expertise, physical (equipment, software, computing facilities), marketing or training related to new technologies or procedures.
- *The Institutional Sector*
  Many of the tasks that a typical firm must perform require co-ordination, either within the firm between various groups of employees or outside it with other suppliers, other firms, and providers of producer services, including finance. There is a variety of ways in which the performance of these tasks can be co-ordinated, each involving different kinds of behaviour. But in general one can distinguish market co-ordination, that relies on the kind of market institutions neo-classical economics usually assumes to be important, and non-market co-ordination that utilises a greater range of institutional arrangements. The latter

depends upon the presence of institutions that regulate the relations between the actors of the system, enhance their innovation capacities and manage conflicts and co-operation. It is possible to distinguish two types of institution (see, for example, Edquist and Johnson 1997): these are a) formal institutions, including employer associations, legal and regulatory frameworks, and b) informal institutions, including the prevailing set of rules, conventions and norms that prescribe behavioural roles and shape expectations.

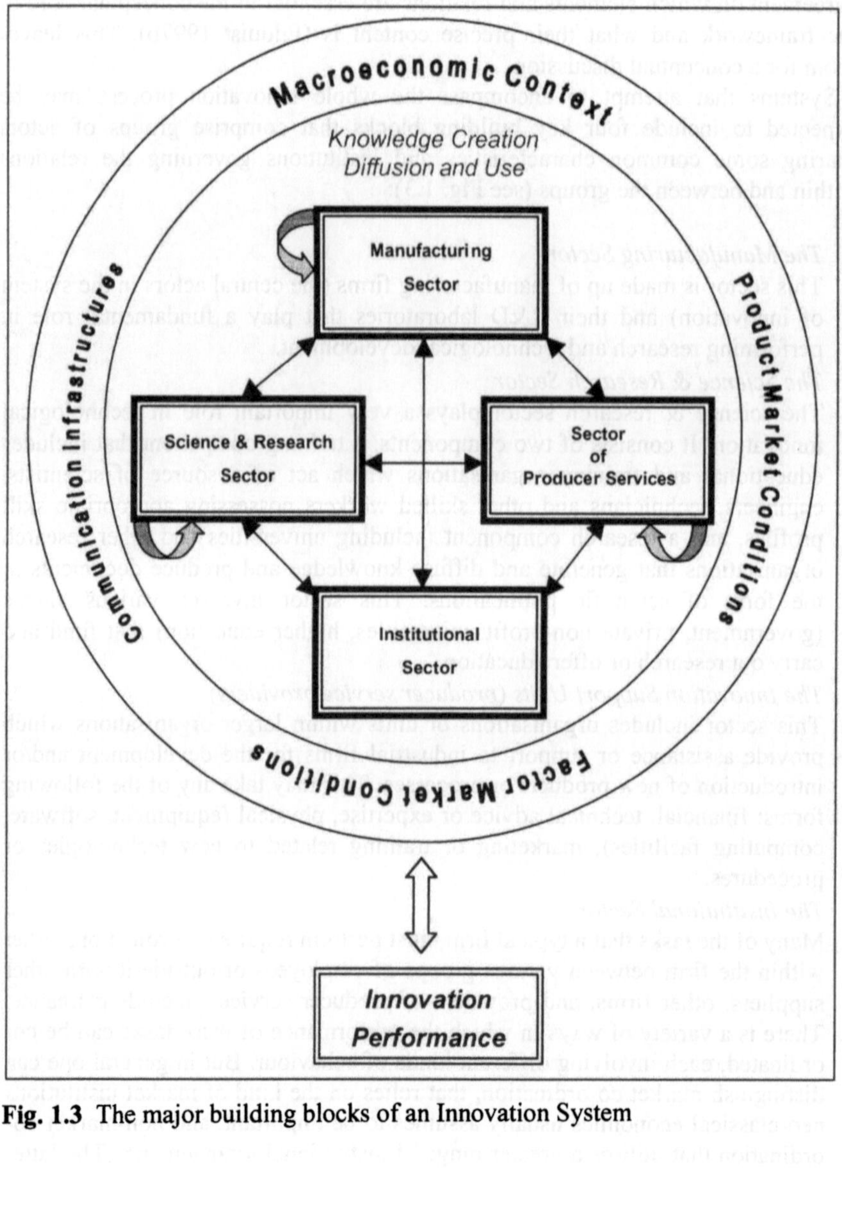

Fig. 1.3 The major building blocks of an Innovation System

To describe and compare systems of innovation in the broad sense, one has to open the boxes of the subsystems, identify the constituent elements and specify the relations in Fig. 1.3 that have importance for innovation performance.

A first source of diversity among systems may be due to differences in the macroeconomic context, the quality of information and communication infrastructures as well as in factor and product market conditions.

The innovation performance of an economy is notably determined by the characteristics and abilities of individual firms and other organisations contingent on its institutions, but it is also very much influenced by the various relations between them, i.e., the ways they interact with each other and with the institutions. The character of these interaction patterns as well as changes in the patterns are central aspects of innovation systems. Linkages within and between the sectors can be specified in terms of flows of knowledge and information, flows of investment funding, flows of authority and labour mobility (scientists, technicians, engineers and other skilled workers), which are important mechanisms for the transfer of tacit forms of knowledge, particularly from the scientific to the manufacturing sector, but also within the latter.

Network analysis may assist in the identification of the central actors in the four subsystems (building blocks) in specific cases, and of the type of information and knowledge they exchange. But there are several unresolved problems. One difficulty is how to define and describe the structure and change of the sector of institutions and to connect it to innovation. The economy's patterns of interaction is made up of different kinds of norms, conventions and established practices that are expected to have important implications for knowledge creation and learning, both inside firms and other organisations, and between them. Searching for and explaining interaction patterns that lead to the creation, dissemination and use of knowledge is part of the systems of innovation approach (Johnson 1997). It can be hypothesised, for example, that the interactions which shape the system will be strong and weak, regular and irregular. This is of course subject to empirical research.

Firms are the main carriers of technological innovation. Their capacity to innovate is partly determined by their own capabilities, and partly by their absorption capacities. Increasing complexity, costs and risks in innovation enhance the role of collaboration and networking in the innovation process to reduce moral hazard and transaction costs. In addition to traditional market-mediated relations, such as the purchase of equipment and licensing of technology, firms exchange information and engage in mutual learning in their roles as customers, suppliers and subcontractors, and even competitors.

A coherent system of innovation necessarily has to include a series of more or less co-ordinated network-like relations such as (Fischer 1999):

- *Customer-producer relations,* i.e. forward linkages of manufacturing firms with distributors, value-added resellers and end users;
- *Producer-manufacturing supplier relations* which include subcontracting, arrangements between a client and its manufacturing suppliers of intermediate production units;

- *Producer-service supplier relations* which include arrangements between a client and its producer service partners [especially computer and related service firms, technical consultants, business and management consultants];
- *Producer network relations* which include all co-production arrangements (bearing to some degree or another on technology) that enable competing producers to pool their production capacities, financial and human resources in order to broaden their product portfolios and geographic coverage;
- *Science-industry collaboration* between universities and industrial firms at various levels pursued to gain rapid access to new scientific and technological knowledge and to benefit from economies of scale in joint R&D, such as direct interactions between particular firms and particular faculty members, or joint research projects, as through consulting arrangements, or mechanisms that tie university or research programmes to groups of firms.

Within the 'systems of innovation' approach to innovation analysis, different types of systems have been defined. A major distinction can be made (see Gregersen and Johnson 1997) between:

- *sectoral* or *technological systems* that are based on the concept of technological regimes and take a specific sector or a specific technology as their point of departure (see, for example, Carlsson 1995; Breschi and Malerba 1997), and
- *localised* (or territorially based) *systems* which are built on some kind of spatial proximity and may be manifested at different geographical scales – as either *local*, *regional* (i.e. subnational), *national* or *global systems of innovation* (see, for example, Lundvall 1992; Nelson 1993; Braczyk, Cooke and Heidenreich 1998; Malecki and Oinas 1999).

Whether a system of innovation should be sectorally/technologically or spatially defined depends on the objective and context of the study being carried out. These two basic variants of the systems of innovation approach complement rather than exclude each other.

1.4 Regional Rather Than National Innovation Systems?

Geographical proximity can be considered as a *necessary, but not sufficient* precondition for the existence of a territorially based system of innovation. A proximity that is only geographic in nature can provide the basis for the presence of an agglomeration of firms, but not necessarily for the presence of an innovation system. The potential of an innovation system depends above all else, as discussed above, on two factors: geographical proximity and technological proximity. *Geographical proximity* indicates the positioning of actors within a given spatial framework, while *technological proximity* pertains to the association with the set of vertical or horizontal interdependencies within the scope of production relationships. The transformation of these two types of proximity into a territorially based system of innovation assumes that they be institutionally organised and structured (Kirat and Lung 1999). Thus, territorially defined systems of innovation are grounded in collective action at a territorial level. The cohesiveness of a territorially based system of innovation is provided by a spectrum of informal institutions, i.e. the set of rules, conventions and norms prevailing at that territorial level (Kirat and Lung 1999).

The concept of territorially based systems of innovation first evolved in a national context (Freeman 1987), and then in a regional context (see, for example, Cooke, Gomez and Etxebarria 1997; Brazcyk, Cooke and Heidenreich 1998; Malecki and Oinas 1999). There is a relatively recently developed tradition of studying national systems of innovation (see, for example, Lundvall 1992; Nelson 1993; Noisi et al. 1993; OECD 1994; Edquist 1997a). Interesting questions and findings have emerged from this literature that sought to establish the extent of convergence and divergence between national innovation systems. This question is of special interest in Europe, given the emergence of innovation-related institutions that have developed simultaneously with European Community institutions (see Caracostas and Soete 1997).

It is increasingly recognised that important elements of the process of innovation have become transnational and global or regional rather than national. The driving forces behind this change are two processes that are simultaneously at work today: the process of globalisation of factor and commodity markets and the regionalisation of knowledge creation and learning. This concurs with the view expressed in Ohmae's work on the 'hollowing-out' of the nation state in an increasingly borderless economic world. He identifies the regional rather than the national level as the economic scale at which leading-edge business and competitiveness is being organised in practical terms (Ohmae 1995). Regions like Baden-Württemberg, Wales, Hongkong-Canton are conceived as much more economically meaningful than, for example, Italy with its abiding north/south divide (Brazcyk, Cooke and Heidenreich 1998).

This awareness does not imply that the national scale is unimportant or irrelevant. This scale continues to be crucial in some circumstances. But it is becoming increasingly clear that there is no *a priori* reason to privilege this particular spatial scale in systems of innovation research, irrespective of time and place (see also Hudson 1999).

A strong case is made today for the growing importance of the regional (i.e. subnational) scale as a mode for innovation systems research. The main argument behind this is that regional agglomerations provide the best context for innovation-based learning economies (Hudson 1999), for knowledge creation, and diffusion and learning. Specific forms of knowledge creation – especially the tacit forms – and of technological learning are both localised and territorially specific. The firms who master knowledge that is not fully codifiable are tied into various kinds of networks with other firms and organisations through localised input-output relations, knowledge spillovers and their untraded interdependencies

(Storper 1997). In some cases market exchange, knowledge spillovers and untraded relations are woven between the various activities within the scope of vertical or horizontal production relationships, but often they are separated.

Formal exchange (i.e. traded interdependencies) and – more importantly – knowledge spillovers and their untraded interdependencies lie at the heart of this line of reasoning:

- *First*, localised input-output relations constitute webs of customer-producer and producer-supplier relations that are essential to communicate information about both technological opportunities and user needs. The user/supplier and producer will gradually develop a common code of communication, making the exchange of information more efficient. To leave a well-established user-producer or producer-supplier relationship therefore becomes costly and involves a loss of information capital (Lundvall 1992).

- *Second*, knowledge spillovers occur because the knowledge created by one firm or organisation is typically not contained within that organisation, but creates value for other firms and other firms' customers. Knowledge spillovers are especially likely to result from basic research, and they are also generated from applied research and technological development. This can occur, for example, in obvious ways such as reverse engineering of products, but also in less obvious ones, such as the abandonment of a particular research line by one firm which signals to others that the line is unproductive and thus saves them the expense of learning this themselves. The spillover beneficiary may use the new knowledge to copy or imitate the commercial products or processes of the innovator, or may use the knowledge as an input to a R&D leading to other new products or processes. Three vehicles of such spillovers may be distinguished: first, the scientific sector with its general scientific and technological knowledge pool, second, the firm-specific knowledge pool and, third, the business-business and industry-university relations that make them possible. Once the key role of knowledge spillovers is recognised, a place for informal institutions appears.

- *Third*, untraded interdependencies or regional assets are less tangible benefits that attach to the process of economic co-ordination and organisational knowledge creation. They are derived from geographical clustering, both economic – such as the development of a pooled labour market – and socio-cultural – such as developed routines, shared values, norms, rules and trust that facilitate interactive processes and mutual understanding in the transmission of information and knowledge. Because tacit knowledge is collective in nature and wedded to its socio-cultural context, it is more territorially and place specific than is generally thought.

Thus, from a more general perspective, it can be argued that it is the combination of territorially embedded Marshallian agglomeration economies, knowledge creation and spillovers and their untraded interdependencies that accounts for the importance of the regional scale in innovation systems research.

There is no doubt that the systems of innovation approach outlined above provides an important framework for understanding why some firms, regions or nation-states are economically successful while others are not. The attractiveness of the systems approach stems from three features which we summarise here:

- *First*, it places innovation and knowledge creation at the very centre of focus, and goes beyond a narrow view of innovation to emphasise its interactive and dynamic nature.

- *Second*, it represents a considerable advance over the network school of innovation (see Håkansson 1987), due to the decisive shift in focus from firm to territory, from the knowledge-creating firm to the knowledge-creating territory.

- *Third*, it views innovation as a social process which is institutionally embedded, and thus lays special emphasis on the institutional context and the forms in which, and through which, the process of knowledge creation and dissemination occurs.

Adoption of this approach overcomes the weaknesses of isolated case studies, because a common conceptual framework is used. Its advantage is that it allows a systematic comparison of innovation activities in different localised systems. Three types of innovation analysis may be performed, depending on the context:

- the first refers to the *micro-level of the system* and attempts to analyse the internal capabilities of selected firms and the links surrounding them (knowledge relationships with other firms and with non-market institutions) with the purpose of identifying unsatisfactory links in the value chain;

- the second refers to the *meso-level of the system* and focuses on specific subsystems and attempts to map knowledge and other interactions within and between subsystems. This may involve the measurement of various types of knowledge flows: a) interactions between manufacturing firms, b) interactions between manufacturing firms and universities including joint research, co-patenting, co-publications and more informal relations, c) interactions between manufacturing firms and other innovation supporting units such as innovation funding and d) personnel mobility focusing on the movement of scientific and technical personnel within the enterprise sector and between the scientific and the enterprise sector;

- the third refers to the *macro-level of the system* and typically involves the use of macro-indicators, such as R&D personnel ratios, R&D expenditure intensity rates, innovation rates, patent intensity rates, and networking indicators of various kinds, which characterise the system in general terms.

1.5 Why Metropolitan Innovation Systems?

There is increasing evidence that metropolitan regions tend to be the principal engines of industrial innovation and growth in national economies. These territories tend to bring about a large share of the outcomes that are considered as the accomplishments of national systems of innovation (see, for example, Oinas and Malecki 1999). Indeed they offer firms spatial, technological, and institutional proximity as well as specific resources whose exploitation generates significant externalities. The opportunities offered by metropolitan regions essentially fall under two headings:

- *Supply of factors of production and infrastructure*: Such factors include the quality of available labour (existence of pools of labour with agglomeration-specific skills and forms of habituation), the availability of capital (for example, the existence of venture capital institutions), communications and research infrastructures (for example, universities and research institutes), or socio-cultural infrastructures that are often critical to the effective operation of the entire economic system.

- *Quality of the regional industrial fabric in terms of subcontractors and suppliers of input*: Full exploitation of technological opportunities requires a satisfactory division of labour between small and large enterprises as well as the co-presence of many different kinds of producers offering specialised inputs and services in timely and flexible response to needs as and when they arise.

Up to now, there have been no studies based on the use of a common rigorous methodology providing evidence across metropolitan regions on questions such as the following: Do firms and other organisations show different patterns of interaction and co-operation? Do firms co-operate more because of policy incentives, for example, between the science and business sectors? How does the institutional set-up affect the ability of firms to innovate? How does proximity (spatial, technological and institutional) affect the development of inter-firm relations in general and innovative collaboration in particular? Will strong network relations internalise knowledge spillovers?

The present book makes a modest attempt to fill this gap in certain aspects. The comparison is based on three metropolitan regions situated at the south western, south eastern and northern periphery of the European Union located away from the core region in Europe: the metropolitan region of Barcelona, the metropolitan region of Vienna and the metropolitan region of Stockholm. They are gateway regions to the South, East and North (Andersson and Andersson 2000). These regions make up an interesting set of study regions, since they are very different in terms of their history, culture and economy. The study nevertheless aims to identify the common features and differences across these regions. In particular, it will seek to provide specific answers about:

(i) variations between metropolitan innovation systems in their innovation capacity and performance within the European Union,
(ii) variations in the structure and changes in the industrial sector in general (including the degree of autonomy and independence of local units in terms of their innovation capacity in relation to their parent company) and customer, producer service/manufacturing supplier and producer network relations in particular,
(iii) variations in the availability and use of technical and innovation support units,
(iv) variations in the structure and changes in the scientific sector and its connection with the business enterprise sector,
(v) the scope, content, and intensity of interregional, national and transnational co-operation of the actors of metropolitan innovation systems.

1.6 Survey Methodology and Design of the Questionnaires

Although there is an increasing amount of empirical research work on innovation processes across Europe in various countries, localities and economic sectors, such research has not generally been carried out in a comprehensive and integrated way. It is thus difficult to address comparisons of metropolitan innovation systems within and across countries of Europe utilising available information. There is therefore an evident need for primary data collection, via postal surveys and/or interviews. It would have been interesting to have combined both methodological approaches, since it would have permitted both quantitative assessments of broad patterns of innovative and networking activities and more qualitative assessments of the context within which the activities are taking place. However, due to limited resources, we had to opt in the present survey in favour of postal surveys.

The innovation picture of metropolitan regions is determined by what happens within and between individual local units in the manufacturing sector, the producer services sector and the science & research sector. Thus, the design of the surveys has focused on individual units in order to identify innovation activity and capability at the metropolitan level. Three postal questionnaires were developed, one for local manufacturing units, one for producer service providers and one for research units (departments of universities, research institutes). The questionnaires were designed in such a way as to encourage high response rates and to ensure consistency of responses. Central control was exercised over the sampling methodologies and the production of the three sets of questionnaires for each of the three metropolitan regions.

The questionnaires used underwent several rounds of development and revision. The survey in the metropolitan region of Vienna were finally conducted from 4 September to 15 December, 1997, in the metropolitan region of Barcelona from

1 October, 1997 to 25 April, 1998, and in the metropolitan region of Stockholm from 1 September to 31 December, 1997.

It is important to recognise explicitly that innovation and networking activities are not independent on the type of industry sector of a local unit. Thus, the questionnaire for local manufacturing units aimed to cover a range of dimensions that likely influence innovative and networking activities, such as the dominant type of product and production processes, capital intensity/labour intensity and firm size. This questionnaire therefore covered the basic local unit characteristics (including the production process), indicators of innovation activities, innovation-related linkages and collaboration. Firms were asked to report on their regional and extra-regional profiles with respect to the location of functions, source and destination of inputs and outputs, and the location of their main competitors and co-operation partners. The reasoning behind such questions was to identify the relative openness of the regions in terms of the spatial organisation of production, externalities, firm rivalry and partnership opportunities. Another set of questions concerned co-operation in innovation in order to identify the extent to which network linkages exist among firms, and between firms and innovation support units and research organisations. This is a way of beginning to explore whether or not and to what extent system-like relationships exist. In addition, questions were posed about the role of the institutional sector, especially formal institutions.

The manufacturing surveys, for example, collected three broad types of data:

- metric data on new product introduction, R&D and other inputs to innovation, sales and employment,
- binary data on, for example, patterns of collaboration and network relations in different stages of the innovation process,
- ordinal data, asking firms to rank the importance of various information sources, obstacles to innovation, support measures, etc.

The questionnaires for the producer service and the science & research sectors were modified to take account of the different types of innovation and linkages that occur in this field.

A wide diversity of organisations and innovation support mechanisms exist in the metropolitan regions under study. This presented a major problem when attempting to classify innovation support units and to measure the nature and extent of their innovation support activities. The feasibility of a postal questionnaire to innovation support units depended on the ability to design a schedule that was sufficiently general to be applicable to a wide range of organisations. Firstly, a clear definition of innovation support units was required. The following working definition has been derived from previous research: an innovation support unit is an organisation, or unit within a larger organisation, that provides assistance or support to local manufacturing units for the development and/or introduction of new products or processes. This may take any of the following forms: financial, technical advice or expertise, physical (equipment, software, computing facilities), marketing or training related to new technologies or procedures.

This definition makes it clear that the innovation support unit may also be a unit within a larger organisation where innovation support or assistance is only a part of their overall activities. Examples of this include departments or units within universities or within commercial accountancy and/or consulting agencies. The key problem concerned the diversity of organisations offering innovation support. This made it difficult to develop a suitable range of questions applicable to every innovation support unit – the organisations were sufficiently diverse for only a very limited range of questions to be appropriate to all. Thus, in line with Fig. 1.3, a decision was made to design specific survey instruments for the sector of producer service providers and for the science & research sector.

The questionnaire for the local units of producer service providers (as defined by NACE 72.1-72.4, 74.2-74.3, 74.4, 74.8 and 74.13-74.14) covered general information (status of unit, age of the unit, number and qualification of employees, turnover). The providers were also asked about the innovation services they offered. These questions were multiple-choice, and aimed to provide an overview of the type and range of support or assistance provided. In addition, further questions attempted to identify the demand for producer services, the industrial sectors served, aspects of the customer/client base, the geographical location of service provision, and the issue of networking with other organisations. Whilst the views of local producer services on attitude (for example, about local barriers and constraints to innovation) are important, it was felt - as in the case of the local manufacturing unit survey – that attitude questions should either be avoided in a postal questionnaire, as they are time consuming for the respondent and do not always produce precise or reliable replies, or handled with great care at the analysis stage.

The final survey served to provide a picture of the general nature of innovation services offered by local units of the science & research sector. The questionnaire was targeted to research departments/units within the following major science fields: architecture, construction, surveying; biology, chemistry, medicine; mathematics, informatics, physics; electrotechnology, mechanical engineering; social sciences including economics and the geosciences. Research establishments were asked not only to provide general information about the nature of the innovation services offered, their customers/clients (including their location) and networking activities with other organisations, but also about the sources of their core funding and other financial resources, as well as the time budget spread over a range of core activities (basic research, applied research, teaching activity, transfer tasks).

Table 1.1 provides evidence concerning population and response rate across both the various sectors and metropolitan regions. Some remarks are important to note in this context.

First, the response rates for the Barcelona surveys were relatively low, and in all cases lower than in the other two metropolitan regions. It is not clear whether this poor outcome reflected difficulties in interpreting or answering the questions or whether other factors contributed. One possible reason is the increasing number of surveys and questionnaires to which firms are asked to respond, another is the nature of the topic. It may be that many smaller firms considered that an investigation about research and innovation did not concern them, but was intended for larger firms with formal R&D activities.

Table 1.1 European metropolitan region innovation surveys: Overall response to the questionnaire surveys

Questionnaire Survey	Metropolitan Innovation System: Vienna	Barcelona	Stockholm
Manufacturing Sector			
Population[a]	908	2,650	1,879
Response Rate[b]	22.5	14.9	24.0
Producer Services Sector			
Population[a]	648	598	1,301
Response Rate[b]	29.3	17.6	25.7
Science & Research Sector			
Population[a]	650	424	346
Response Rate[b]	44.6	34.9	50.0

Notes: **a** total number of corresponding local units adjusted for closures, untraceable units, establishments no longer engaged in manufacturing etc. and thus no longer appropriate to the survey.
b number of corresponding local units divided by the total number multiplied by 100.

Second, bias in postal questionnaires can be a serious problem, particularly when the response is as low as in the Barcelona metropolitan region. Evidence for bias in terms of the key variables of interest is not easy to find as – by definition – the information is not known. Non-response in terms of non-return of questionnaires is the principal problem. Telephone-based surveys of small subsamples of non-respondents have been conducted in the case of the manufacturing survey to analyse differences between local units that responded unprompted to the postal surveys and those that responded during the telephone chase-up. No significant differences in terms of innovative activity were revealed in the metropolitan region of Vienna once the effects of local unit size had been taken into account. By contrast, the Barcelona survey indicates evidence for bias affecting the key variables of interest. There are implications arising from the bias towards larger local units, on the one hand, and towards high tech industrial sectors, on the other. This means that the Barcelona innovation survey tends to overstate aggregate levels of innovative and networking activities within the population. This bias had to be taken into account in making the comparisons.

Third, it is important to note that postal questionnaires for producer service providers in the metropolitan region of Barcelona were sent out only to firms with at least 10 employees. This has to be taken into account in the comparative part of the analysis.

Finally, it should be taken into account that the metropolitan regions differ in size due to different practical and political criteria used for the spatial delineation. This is reflected also in the different sizes of the survey populations and will have implications on the comparative stage of the analysis, especially with respect to the issue of cross-regional networking activities.

1.7 A Brief Guide to This Volume

The three empirical metropolitan studies represent the heart of this volume. They are unified by a single conceptual framework and survey methodology based on centralised control and design of near-identical survey instruments in order to allow direct comparisons between the various regions.

Armed with the results from the questionnaire survey of manufacturing local units, producer service providers and research establishments, the chapters that follow aim to shed some new light on the innovation and networking activities in the regions concerned. To facilitate comparison, the authors have developed a relatively detailed list of items all metropolitan chapters should cover. Chapter 2 focuses on the metropolitan region of Vienna, Chapter 3 the metropolitan region of Stockholm and Chapter 4 the metropolitan region of Barcelona. In addition, the Annex provides further detailed empirical data in form of numerous tables. Chapter 5 then attempts to summarise some areas of similarity between the metropolitan systems and points to some striking differences as well. The development paths of the three metropolitan regions have been very different and so too is the present-day organisation of industry and the structure of their innovation systems. The reasons for these differences reside to a significant degree in differences in the national and metropolitan histories and cultures. These have profoundly shaped institutions and policies. Exactly how, will be discussed in the chapters that follow.

2 The Vienna Metropolitan System of Innovation

In this chapter we shall discuss the anatomy of the Vienna innovation system. We begin with a profile of the region concerned (Section 2.1) followed by a brief characterisation of the Austrian R&D system (Section 2.2) and then move on to the institutional setting of the innovation system in order to illustrate the institutions and mechanisms supporting technological innovation (Section 2.3).

Public money generally supports not only research at universities and public research establishments, but also R&D in industry. But this tends to vary across nation-states and regions. These differences are an important issue of the comparative study. As we already have stressed, technological innovation involves much more than R&D. The majority of firms, not being large, probably do not have R&D budgets, but may nevertheless be innovators. They tend to be incremental rather than radical innovators, operating in markets where innovation is orchestrated through supply chains in relatively mature products. Thus – as we stated earlier – innovation is often a continuing process, with product and process engineers learning from experience and making modifications on that basis, and customers feeding back complaints and suggestions.

Armed with the results from the questionnaire survey of manufacturing local units, Section 2.4 sheds some new light on the innovation and networking activities in the Vienna metropolitan innovation system, while Section 2.5 turns attention to the sector of producer service firms that provide assistance or support to industrial local units for the development and/or introduction of new products or processes. Because technological innovation is increasingly an interactive process involving specialist technological requirements and users who are very demanding, far more needs to be done to increase the possibilities of firms to engage in co-operative interaction in pursuit of innovation, while improving their competitiveness. Of particular importance here is the process of exploiting knowledge for commercial gain. Thus, sources of knowledge capital created in the science & research sector must be brought closer to the business world without compromising the important scholastic and critical functions of universities and non-university based research establishments in an innovation system.

Based on a large scale empirical survey of the research establishments in the metropolitan region of Vienna, Section 2.5 provides evidence on the structure of the science & research sector and its linkages with the industry sector. Sections 2.4 to 2.5 present information on customers, manufacturing suppliers, producer service providers and producer network relations, as well as the relations between industry-university-research establishments of manufacturing firms located in the metropolitan region of Vienna, while Section 2.6 examines the question of the extent to which the likelihood of networking is influenced by firm specific attributes, as suggested by the resource-based view of the firm. The major results are summarised and some conclusions are drawn in the final section.

2.1 The Metropolitan Region

During the 20th century, Austria's position in Europe changed several times. Each change affected regional development in important ways. This evolution also had a strong impact upon Vienna. As a city of 2.1 million inhabitants, before World War I it was the rapidly growing capital of the Austro-Hungarian Empire. At that time, this was a state with a population of more than 50 million and a strong orientation towards economic activities in Eastern and South-eastern Europe. Vienna, having played for centuries an essential role in the history of Europe and the world, suddenly lost that role at the beginning of the last century to become the capital of the Republic of Austria, a small country with currently only 8 million inhabitants.

The first decade of the new millennium will pose an extraordinary challenge to the European Union in general and to the metropolitan region of Vienna in particular. No matter whether the EU enlargement comes – in five, seven or ten years' time – it will set new political, economic and social standards in European development that already today call for preparatory steps to be taken on both sides of the EU border. In any case, with the projected EU enlargement, Vienna will become the centre of a new and exciting European cross-border region. The Vienna City Government has already started specific initiatives in order to meet the challenges and opportunities presented by this development, in co-operation with the neighbouring provinces of Lower Austria and Burgenland, as well as with regions across the border including the capital city of Slovakia, Bratislava. Joint efforts to improve the transportation and telecommunications infrastructures and to attract international investors will strengthen the competitiveness of the entire region – and thus also the economic power of its core.

The metropolitan region of Vienna is located at the Eastern extreme of Austria and the EU – Bratislava being only 50 km from Vienna (see Fig. 2.1). With its 1.6 million inhabitants in 1991 and a workforce of 830 thousand, the city of Vienna forms the core of the metropolitan region (2.1 million inhabitants and a labour force of 1.09 million). Over the period 1981-1991, industrial employment declined from 21 percent to 16 percent of the workforce, while services rose from 30 percent to 38 percent. In 1991 the rate of unemployment was 4.6 percent in the metropolitan area. GRP per capita is approximately 29,000 Euros, the most important industries are electronics, transportation, construction, chemicals and fabricated metals.

As a political unit, Vienna is not only the largest municipality in Austria, but is one of the nine autonomous provinces in the federal system. Thus, the exercise of provincial and municipal governmental authorities coincide. As a result, Vienna has significant capabilities for promoting regional economic development and innovation. Promotion policies are part of the social partnership arrangements (see Section 2.3 for more details on the institutional set-up). Some political tensions do exist with the districts that constitute the outer areas of the metropolitan region. These belong to the province of Lower Austria. In alphabetical order the districts are as follows: Baden, Bruck an der Leitha, Gänserndorf, Korneuburg, Mödling, Tulln, and Wien Umgebung. This political division of the metropolitan region into two separate provinces and several communities thwarts a clear-cut and comprehensive policy for the development of the metropolitan region.

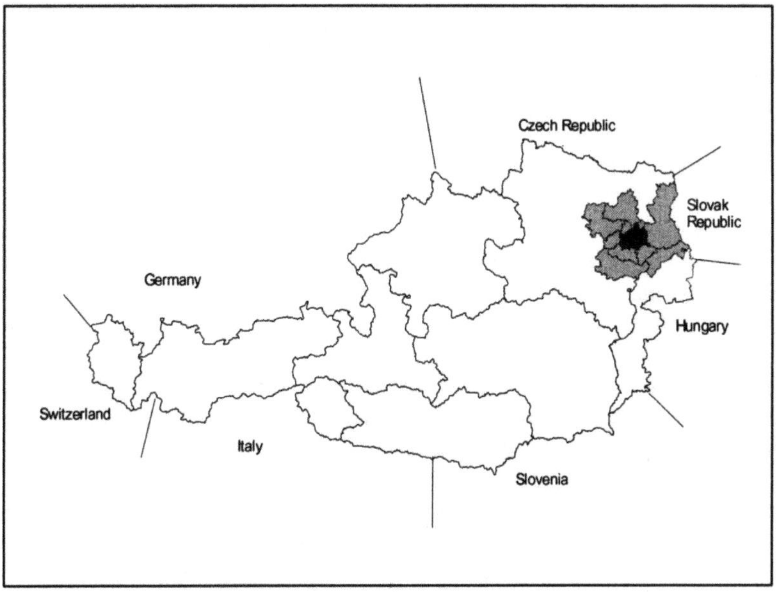

Fig. 2.1 The Vienna metropolitan region

2.2 Some Features of the Austrian National Innovation System

The purpose of this section is to briefly characterise some aspects of the Austrian R&D system. An R&D system is at best a poor proxy to an innovation system, but since R&D information is the only reasonably coherent and comprehensive data available, there is no choice but to use it to give some picture of the national system within which the metropolitan system is embedded. We do not intend in this section to give a detailed description of the institutions and their activities in the field of R&D and public technology in Austria, but will make an effort to highlight the most important characteristics of the R&D system.

Formal, reported R&D expenditures are only a part of the innovation-related outlays made by firms. Such data ignore the complex processes of technological accumulation whereby tacit knowledge is built up and transferred from one generation to the next within firms and organisations. R&D captures nothing of the linkages between organisations, the feedback processes or the relationship between firms and agencies.

The overall structure of the R&D system can be approached by examining the main aggregates in terms of funding and execution as displayed in Fig. 2.2 for the year 1993. The main structural features may be summarised as follows:

- The public sector (i.e. the federal government, provincial and municipal governments, chambers of commerce, social insurance organisations, etc.) funds about 45 percent of R&D, the business enterprise sector approximately 53 percent, and the rest coming from foreign sources.
- About 57 percent of R&D is executed in the business enterprise sector and 43 percent in the science & research sector.

The way in which the official budget documents and other reports are presented makes it extremely difficult to understand the precise pattern of allocation of funds. Without going into details, the following observations should help to give a fuller understanding of the data:

(i) The *science & research sector* is divided into three segments: the higher education segment, the government non-university research segment and the private non-profit research segment.

(ii) The *higher education sector* does not include only universities and art schools (i.e. colleges of university status), but also the Austrian Academy of Sciences, the experimental units in higher technical schools and university clinics.

(iii) The *government non-university research sector* includes federal, provincial and municipal government R&D laboratories, and also institutions of professional associations, as well as social insurance institutions, museums and public hospitals.

(iv) The *private non-profit research* sector includes private non-profit organisations, such as the Ludwig Boltzmann Society. These are mainly of a private or civil, denominational or another non-public status.

(v) Finally, it is important to note that some research institutions, such as the Austrian Research Centre Seibersdorf, are positioned in the business enterprise rather than the science & research sector if they are organised under company law and have turnover proceeds from contract research and project funding schemes exceeding 50 percent of the overall total.

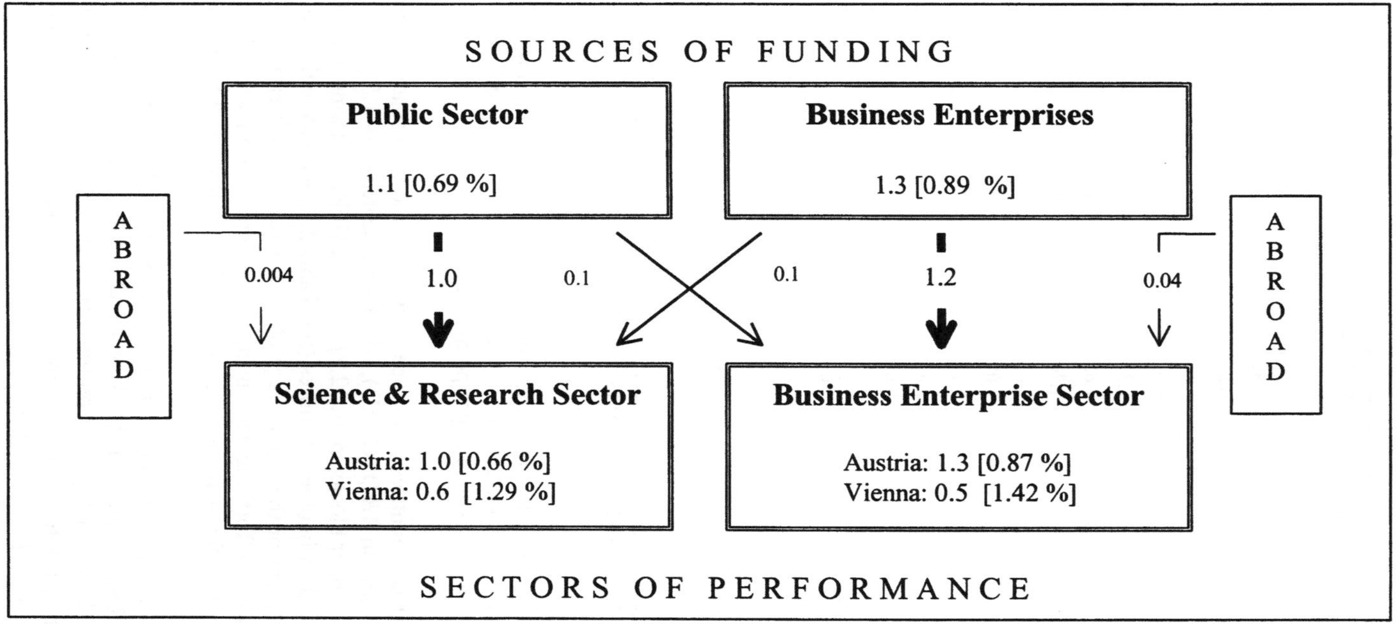

Fig. 2.2 Sources of funding and performance of sectors (1993)
(Figures in Euro billion, those in brackets represent GERD in terms of GDP or GRP)

Table 2.1 Ratio of GERD to GDP in selected countries (1991, 1993, 1995)

	1991	1993	1995
Austria	1.50	1.49	1.53
Sweden	2.89	3.28	3.02
Spain	0.87	0.91	0.80
Ireland	0.96	1.21	1.40
Italy	1.32	1.26	1.14
Norway	1.65	1.73	1.59
Denmark	1.70	1.79	1.82
Netherlands	2.05	2.00	2.04
Finland	2.07	2.21	2.32
United Kingdom	2.11	2.15	2.05
France	2.41	2.45	2.34
Germany	2.61	2.43	2.28
EU	1.96	1.94	1.84
Japan	3.00	2.88	3.00
USA	2.84	2.64	2.58
OECD	2.31	2.20	2.16

Source: OECD, quoted in Hutschenreiter et al. (1998).

These observations have to be taken into account when considering Fig. 2.2 in greater detail. From a closer look at the available general data and in particular those summarised in Fig. 2.2, seven major points emerge:

- Among the industrial countries of Europe, Austria is a lower spender of gross domestic expenditure on R&D (GERD), which represented only 1.5 percent of GDP in 1993 (the upper blocks in Fig. 2.2). Clearly, GERD in terms of GDP is smaller in Austria than in other countries (see Table 2.1) and below the OECD average of 2.2 percent or the EU average of 1.94 percent. But it is interesting to note that the R&D efforts carried out in the metropolitan region of Vienna are much higher than for Austria as a whole – 2.7 percent of gross regional product (GRP) in 1993.
- It is the business enterprise sector that plays a less important role in funding and performing R&D. Industrial R&D, or R&D carried out within firms (see lower right-hand block in Fig. 2.2) represented only 0.87 percent of GDP in 1993. This is far below the OECD average of 1.46 percent (Hutschenreiter et al. 1998). One reason for this gap is certainly the large share of small firms

which focus on labour-intensive, low-technology or medium-technology production sectors. But it also reflects distortions in the allocation of capital and insufficient availability of R&D and production skills (OECD 1995b).

- The share of public funds in industrial R&D expenditures – in the form of direct research and technology subsidies – was about 9 percent in 1993. The role of public funding is therefore not very different from other countries, especially if defence-related expenditures are excluded.
- About 43 percent of the R&D activities (measured in terms of GERD) took place within the science & research sector, i.e. one-third in the higher education sector as defined above (34 percent), the rest in public non-university research establishments (9 percent) and private non-profit research establishments (much less than one percent).
- Federal government traditionally concentrates its financial support on R&D capacities lodged within the higher education sector, with its dual structure of universities and the research establishments of the Austrian Academy of Sciences. More than two thirds of public R&D funds were allocated to this component of R&D, but this figure narrows down considerably when it is taken into account that the general university funds provided are basically for teaching and only subordinately for research.
- The linkage of the business enterprise sector with the science & research sector is weak in terms of flows of funds. This seems to be similar to the situation in Sweden and Germany, but in contrast to Spain, the UK and the Netherlands (Hutschenreiter et al. 1998). Only about 7 percent of R&D performed within the science & research sector was financed by the business enterprise sector in 1993. The more entrepreneurial among the heads of university institutes negotiate with outside sources in the business enterprise sector for the extra funds needed to develop high quality research. However, they often remain hampered in their efforts to establish R&D contracts with industry. Discussion in recent years about the need for closer university-industry relationships has not yet freed university research from the fetters of traditional state accounting procedures.
- According to the statistics summarised in Fig. 2.2 about two percent of R&D was financed from foreign resources in 1993. This figure reflects a low level of internationalisation of R&D, but was also influenced by the fact that Austria was not at that time a member state of the EU. Recent figures on the Austrian participation in the Fourth Framework Programme of the EU reveal that this picture has been changing rapidly in recent years.

2.3 The Institutional Set-Up

The tasks a typical firm has to perform require co-ordination either within the firm between various groups of employees or outside, with other firms, suppliers and service providers. There are various ways in which the performance of these tasks may be co-ordinated, each inducing a different kind of behaviour and leading to a different outcome. One can distinguish between market co-ordination and non-market co-ordination. The latter depends upon the presence of institutions which provide the actors in the system with opportunities for negotiating, monitoring and enforcing agreements between each other. A variety of institutions can provide such capacities, including associations, the chambers of labour and trade unions, the federal and provincial administrations, and various frameworks for standard setting. Effective non-market co-ordination often depends upon particular combinations of institutions, such as those that superintend wage bargaining in many nation-states. These institutions and the co-ordination they support tend to play a special role in five arenas external to the firm: the negotiation of wages and working conditions, vocational training, corporate governance and finance, and supplier relations. But such co-ordination among firms may be also found in an increasing range of spheres, including research, product development, marketing, and joint production arrangements. Non-market co-ordination in the political economy is not only important outside the firm, but also inside. There tend to be systematic differences across nation-states in forms of internal co-ordination, which correspond to differences in the kinds of co-ordination present in the environment external to the firm. The presence of institutional complementarities of this kind is a key aspect of co-ordination. Effective co-ordination in any of the spheres mentioned above often requires the presence of multiple institutions that interact in such a manner as to reinforce the advantages that the other institutions offer to the firms (Hall 1997).

It is evident that effective non-market co-ordination is not easy to establish. Since the institutions that support it provide collective goods, individual firms generally do not have the incentive or means of establishing them – the state is better positioned to do so. Also, because effective co-operation usually demands a great deal of information sharing and high levels of trust among firms, the presence of a supporting institutional framework alone is rarely sufficient to persuade firms to co-operate with each other. Common experience of successful co-ordination in at least some of the spheres mentioned above [for example, negotiation of wages and working conditions, vocational training] can be a prerequisite for further co-ordination in others. The tendency of firms to distrust the state and its agencies renders it particularly difficult for public authorities to construct and establish such arrangements. Indeed, particular forms of co-ordination among and within firms tend to develop over long periods of time as firms gradually adjust their behaviour accordingly (Hall 1997). As it is primarily the nation-state that provides the legal regulations on which many forms of co-ordination depend and within which the institutions supporting co-ordination have evolved, systematic differences in forms of co-ordination and firm behaviour tend to be found across nation-states and, thus, between their metropolitan regions. This means that the capacity of firms in one particular metropolitan region to pursue various kinds of endeavour will be conditioned by the types of inter-firm and intra-firm co-ordination that are available. These in turn depend on the presence of institutions supporting such co-ordination.

Co-ordinated market economies – in contrast to liberal market economies – are characterised by corporate governance institutions which facilitate the exchange of information among firms via relatively powerful supervisory boards, cross-shareholding, or the extensive involvement of suppliers of finance in corporate finance. This tends to restrict the discretionary powers of chief executive officers, but provides financial sources that do not depend on share-price. This reduces the extent to which firms must remain oriented toward short term profitability, and favours securing agreement to corporate strategies from a variety of corporate actors. Such conditions make long term employment relations more feasible, putting a premium on securing labour peace. Accordingly, firms in such institutional settings are more likely to accept works councils and powerful trade unions, provided that forms of effective co-ordination can be established. This is more feasible in such settings because close relations with other firms make it possible to develop sectoral or national arrangements for wage co-ordination that can in turn be extended to such spheres as vocational training (Hall 1997).

These institutional arrangements are not unique to Austria. A number of other nation-states with political and economic systems described as 'consociational' or corporatist (like Germany and Japan) have similar institutions. But they stand in contrast to the institutional features of other countries, such as Britain and the USA, where decision-making in both the economic and political arenas is organised on a more hierarchical basis which concentrates power in the hands of a few actors.

Unique to the Austrian system is the chamber system that forms the framework of the social partnership. This system may be briefly described as follows. On both sides of the labour market, there is a parallel set of voluntary organisations such as industrial associations, trade unions etc. and self-governing bodies called chambers. Membership of the chambers is compulsory, and the chambers are financed basically through contributions related to the wage bill. The chambers on each side of the labour market are hierarchically organised, with two central chambers for workers and employers, respectively. The institutional centrepiece of social partnership is the Joint Commission for wage and price issues where, in addition to the central chambers, the Federal Government and the Federation of Trade Unions are also represented. Four subcommittees are responsible for the centralised surveillance of sectoral wage agreements, price developments and competition policy, wider issues of a social and economic character, and international issues, respectively. Within this institutional set-up, the chambers represent their members vis-à-vis the legislative and administrative powers. They have the right to present comments on government draft bills and furthermore are represented in many institutions. This implies that the social partners, or more precisely their chamber representatives, have a decisive influence on many aspects of policy (OECD 1995b).

This kind of institutional setting provides the firms located in the metropolitan region of Vienna with advantages for pursuing certain kinds of endeavour. Long term employment and closer relations with the labour force make it easier for them to pursue strategies based on *incremental* product or process innovations of the kind that builds on the knowledge embodied in the labour force. The presence of institutions capable of co-ordinating vocational training and pressure from powerful trade unions favours forms of production based on high cost and highly skilled labour focusing on high value added product lines. In the coming years, rising rates of growth in Central East Europe should provide new markets for such products. Conversely, such a setting tends to militate against radical innovation in new product lines, because it can be difficult to recruit expertise from outside the firm and to change the skill categories of the labour force. Radical changes in corporate strategy involving reorganisation or reorientation of the labour force may be difficult because the discretion of chief executive officers is limited by more influential supervisory boards and/or works councils.

2.4 The Manufacturing Sector

For several reasons, the problems closely connected to the set of processes often summed up by the term globalisation are more intense for the metropolitan region of Vienna than for the other metropolitan regions. First, the collapse of communism in Eastern Europe meant the collapse of what once had been, in effect, substantial trade barriers. Firms in the Vienna region now sit closer than those of many other regions to large, low-cost labour markets that in some cases provide highly attractive alternative sites for production. Second, due to high levels of productivity, Austrian labour costs are now among the highest in the world, thereby intensifying fears that Austrian products may be undercut by foreign competition. Under these conditions – reinforced by EU membership – the future contains important new challenges for the manufacturing sector. In this context, it is inevitable that some firms will decide to move parts of their production abroad and that others will face more intense competition from foreign firms. However, in the long run, the growth of foreign economies, especially in East Central Europe, should provide new markets for Austrian goods.

A The Survey

This section is based on the survey carried out in the metropolitan region of Vienna, and looks at manufacturing firms, the central actors in the metropolitan innovation system, and their R&D laboratories, which play a fundamental role in generating new technologies and products. Data were collected from the population of 908 manufacturing firms with at least 20 employees, as identified by the Firm and Product Database Register for 1995 organised and managed by the Department for Systems Research at the Austrian Research Centre Seibersdorf. 204 firms returned the completed questionnaire, resulting in a response rate of approximately 22.5 percent. This figure was not as high as expected, but still acceptable given the conditions under which the survey was carried out, and much higher than in the case of the metropolitan region of Barcelona.

Table 2.2 Response patterns and response rate of responding manufacturers

	Total Number of Registered Firms 1995		Number of Responding Firms 1997		Response Rate [a]
	no.	%	no.	%	
Industry Sector					
Textiles & Clothing	72	7.93	13	6.37	18.05
Food Industry	112	12.33	24	11.76	21.43
Wood, Paper & Printing	198	21.81	49	24.02	24.75
Chemicals, Plastics & Rubber	185	20.37	38	18.63	20.54
Electrical & Optical Equipment	115	12.67	28	13.73	24.35
Basic Metals & Metal Products	108	11.89	24	11.76	22.22
Machinery & Transport	118	13.00	28	13.73	23.73
Employment Size					
≤ 49	396	43.61	88	43.14	22.22
50 – 99	225	24.78	49	24.02	21.78
100 – 499	232	25.55	54	26.47	23.28
≥ 500	55	6.06	13	6.37	23.64
Total	908	100.00	204	100.00	22.47

Note: a number of responding manufacturing firms divided by total number of firms multiplied by 100.

Table 2.2 presents a breakdown of the sample responses and illustrates the response rates for seven manufacturing sectors and for four firm size classes as measured by employment. Using the standard NACE classification sample firms were allocated to the following industry sectors: textiles, clothing and leather (NACE 17-19); food, beverages and tobacco (NACE 15-16); wood, paper and printing (NACE 20-22, 36); chemicals, rubber and plastics (NACE 23-26); electrical and optical equipment (NACE 30-33); basic metals and metal products (NACE 27, 28, 37); machinery, transport equipment (NACE 29, 34, 35).

The sample can be seen to broadly reflect the overall structure of the total population. As explained, the lower response rate by small local manufacturing units may be attributed to the fact that such firms are less likely to undertake any kind of formal R&D activity, since they tend to lack the resources. In some cases they tended to dismiss the questionnaire as irrelevant to their circumstances. This problem is however general and not specific to the Vienna survey (and was found to be more pronounced in other surveys). A telephone-based survey of a small sub-sample of 90 non-respondents indicated that it did not have a significant impact on the results.

The majority of surveyed firms were small, 64.7 percent had less than 100 employees, compared to 68.4 percent of the identified population, and many of these (49.6 percent of those with a known starting year) have been in business since 1970. In terms of organisational status, 111 firms (55 percent) were independent, the remainder operated as a main plant (36.1 percent) or as a branch plant (8.9 percent) within a larger parent group

B Innovation Activities of Manufacturing Firms

Table 2.3 shows a brief profile of the surveyed firms utilising five indicators. The first three attempt to capture the resources to which the manufacturing firms have access for the purpose of innovation:

- the presence of continuous on-site R&D facilities,
- R&D employment in terms of the R&D personnel ratio,
- R&D expenditure in terms of the R&D expenditure intensity (as percentage of sales turnover).

A further set of two indicators focuses on innovation activities or output and includes:

- the actual introduction of new products (averaged over 1994-1996) per 1,000 employees, i.e. the product innovation rate,
- the share of turnover accounted for by new or improved products (averaged over 1994-1996).

Table 2.3 Selected characteristics of surveyed firms (1994-1996)

	Firms with Continuous On-site R&D 1997	R&D Personnel Ratio [a]	R&D Expend. Intensity	Innovation Rate [b]	% of Turn-over by Product Innovation
Industry Sector	c				
Textiles & Clothing	2 (15.38 %)	17.76	5.84	62.53	0.13
Food Industry	3 (12.50 %)	28.18	1.76	34.02	0.28
Wood, Paper & Printing	4 (8.16 %)	11.50	1.55	27.71	0.04
Chemicals	5 (13.16 %)	53.29	6.39	41.84	0.19
Electrical & Optical Equip.	7 (25.00 %)	250.41	16.05	6.15	0.51
Basic Metals & Metal Prods.	2 (8.33 %)	26.18	2.30	11.99	0.53
Machinery & Transport	7 (25.00 %)	25.50	5.21	4.01	0.50
Employment Size					
≤ 49	7 (7.95 %)	51.74	2.84	128.12	0.13
50 – 99	7 (14.29 %)	30.54	3.18	86.83	0.16
100 – 499	11 (20.37 %)	32.18	4.35	6.51	0.32
≥ 500	5 (38.46 %)	142.59	10.05	2.46	0.44
Production Size					
Custom Production	11 (12.09 %)	39.03	4.56	27.75	0.26
Batch Production	6 (10.71 %)	176.21	11.40	15.73	0.42
Custom & Batch Production	1 (12.50 %)	37.75	2.92	34.82	0.12
Mass Production	10 (29.41 %)	66.64	6.81	6.96	0.27

Notes: **a** per 1,000 employees;
b denotes number of new products per 1,000 employees;
c percentage of all firms of the corresponding category.

The second of these measures is an indicator favoured by many management experts as a measure of a firm's innovative capacity and is a widely accepted measure in the benchmarking literature (see, for example, Zairi 1992). It relates product innovations to economic activity. It is accepted that the definition of what constitutes a new or improved product is problematic and this has to be taken into account when considering the figures given in Table 2.2. In some industry sectors, such as the food industry and textiles and clothing, new or more particularly 'improved' products may appear rapidly, while in others four or five year development cycles may be the norm. In sectors such as machinery and transport, for example, very long lead times are still the case.

Following Malecki and Veldhoen (1993), we classified firms as innovative if they met the following criterion: that product innovations introduced during the past three years comprised more than 20 percent of the firm's annual turnover. Defined in this way, only 50 (26.5 percent) of the firms can be considered as innovative. 64 percent of these had fewer than 100 employees and 16 percent had under 50 employees. The sectoral distribution indicates a predominance of innovative firms in the categories: electrical and optical equipment, machinery and transport and basic metals and metal products. These three sectors account for 50 percent of all innovative firms. Of the non-innovative firms, 45.3 percent are engaged primarily in custom production, 26.6 percent in batch production and another 5.0 percent in custom and batch production. This suggests that flexible production, particularly of custom products for individual customers, is the norm rather than the exception among the firms surveyed, whether or not the concept of 'new/improved' products is appropriate.

Those establishments without continuous on-site R&D generally fell into two categories: small independent enterprises and the branches of group organisations. Some independent enterprises were carrying out subcontract work and were thus dependent upon customer specifications rather than original research, still others carried out research on an *ad hoc* basis where and when necessary. But the most important explanations for the lack of on-site R&D tended to be the small size of the enterprise or, in the case of branch plants, the corporate decision that centralised R&D facilities were adequate for the needs of the establishment.

While continuous on-site R&D indicates commitment to inventive and innovative activity, the size of the R&D facility measured in terms of employment and/or finance can give some indication of the scale of this commitment. Establishments with higher scales of R&D expenditure and R&D employment were shown to be much more likely to be innovative than those with lower. This variation needs to be taken into account when considering differences between manufacturing firms in the incidence of innovation activities. This can be done by means of a logit analysis. Fig. 2.3 thus shows that, taking a domestically owned multiple plant company as a benchmark, an increase in R&D expenditure and R&D personnel are the factors most likely to increase the success of innovation activities. These were measured in terms of the share of turnover accounted for by new or improved products averaged over 1994-1996. The organisational status of an independent plant is likely to significantly affect the chance of successful innovation activities irrespective of establishment size. Employment size and ownership do not have a statistically significant influence.

C Networks and Network Formation

R&D may be misleading or at least incomplete as an indicator of technological capability, because it does not include network activities, learning, informal R&D and other means of enhancing a firm's knowledge base (Malecki 1997). The performance of a firm may be best viewed as a product of the interplay between in-house R&D innovation efforts and external innovation networks for knowledge transfer. The knowledge needed to compete comes most often from customers, suppliers (manufacturing and producer service suppliers) and from other firms and

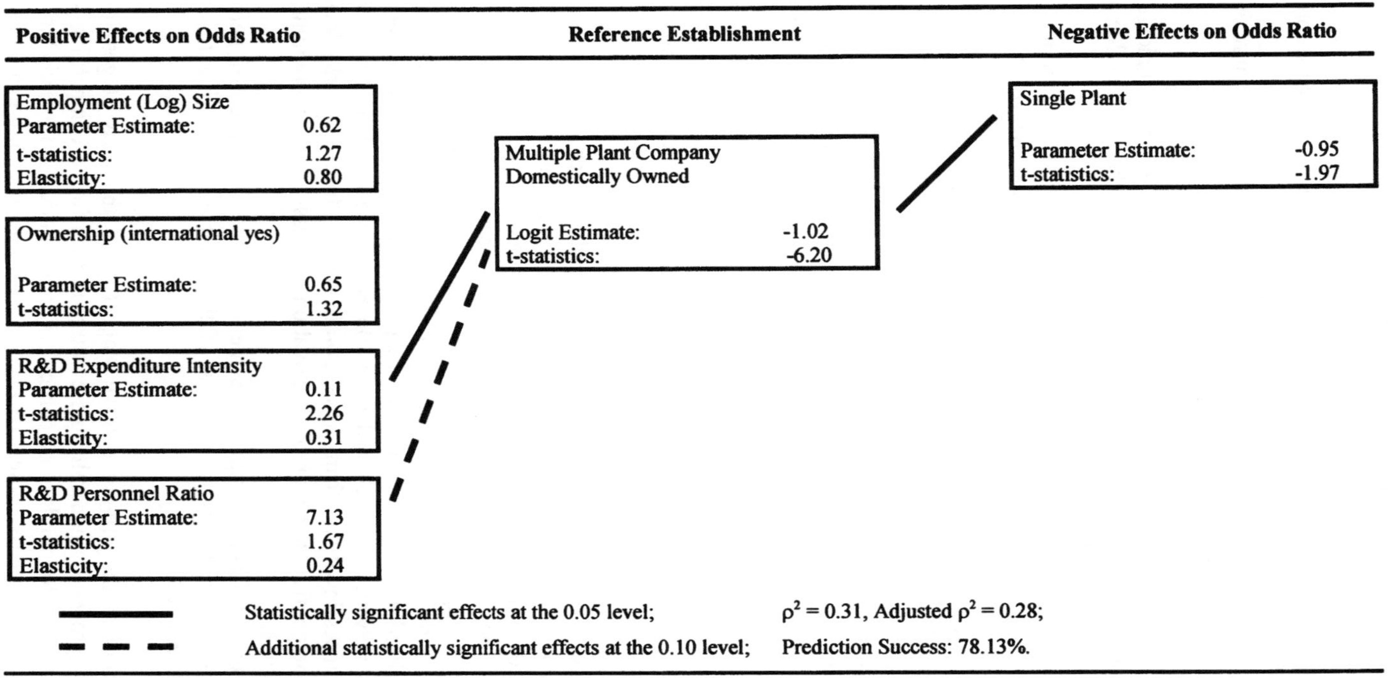

Fig. 2.3 Success of innovative activities measured in terms of share of turnover accounted for by new or improved products

institutions. The innovative capacity supported by regional inter-firm networks not only supports existing firms, but also offers opportunities to open up new businesses in order to serve newly identified markets. The importance of networks and of innovative niches in sparking innovation applies in both high-technology industries and in traditional sectors.

Network activities of manufacturing firms in the Vienna metropolitan region are organised around five types of network:

- *customer networks,* which are defined as the forward linkages of manufacturing firms with distributors, marketing channels, value-added resellers and end users,
- *manufacturing supplier networks,* which include subcontracting, arrangements between a client (the focal manufacturing firm) and the manufacturing suppliers of intermediate production inputs,
- *producer service supplier networks,* which include arrangements between a client (the focal manufacturing firm) and its producer service partners (especially computer and related service firms, technical consultants, business and management consultants, market research and advertising),
- *producer networks,* which include all co-production arrangements (bearing to some degree on technology) that enable competing producers to pool their production capacities, financial and human resources in order to broaden their product portfolios and geographic coverage,
- *co-operation with research institutes/departments of universities* (pre-competitive stage) pursued to gain rapid access to new scientific and technological knowledge and benefit from economies of scale in joint R&D.

Firms pursue co-operative arrangements of this kind in order to tap into sources of know-how located outside the boundaries of the firm, gain fast access to new technologies or new markets, benefit from economies of scale in joint R&D and/or production, and share the risks for activities that are beyond the scope or capabilities of a single firm. The picture which emerges from the evidence of the current study is that there exists a maze of different networks. They range from highly formalised to informal network relations, from highly specialised and rather narrow networks to much wider and looser networks such as, for example, technical alliances involving firms as corporate entities, from networks focusing on the pre-competitive stage of the innovation process to those involving the competitive stage.

Table 2.4 provides some empirical evidence on the five types of networks described above, from the point of view of the focal manufacturing firm, and highlights the fact that:

- Co-operation in the pre-competitive stage (i.e. in the early stages) of the innovation process is generally more common than in the competitive stage. External information tends to be particularly relevant during the early stages of the innovation process, when perception of problems and evaluations of technological possibilities take place.

- Customer and user-producer (i.e. manufacturing and producer service supplier) relationships are much more frequent than horizontal forms of co-operation such as producer networks and research institution-industry linkages. Producer network linkages are typically found among smaller enterprises that may occasionally also be competitors. Customer networks represent the most frequent form of inter-firm co-operation – activities with customers and suppliers constituting 35.3 percent of all such activities. Manufacturing and producer service suppliers have strong incentives to establish close relationships with user firms and even monitor some aspects of their activity. Knowledge produced as a result of learning-by-using can only be transformed into new products if the producers have direct contact with users. In turn, user firms will generally need information about new products or components. This may not only mean awareness, but also quite specific inside information about the way in which the new, user-value characteristics relate to their specific needs.
- 37.7 percent of manufacturing firms were integrated into customer networks, 27.9 percent into manufacturing supplier networks, 46.6 percent into producer service supplier networks, and only 18.6 percent have set up co-operative relations with research institutes and/or departments of universities, despite the active promotion of university-industry programmes in Austria.

As in other studies (see, for example, Meyer-Krahmer 1985) three clusters of manufacturing firms may be distinguished. The first cluster, characterised by a strong outward orientation, frequently utilises the whole range of possibilities in obtaining external knowledge. Firms in this cluster share widespread network activities, in both the pre-competitive and the competitive stage of the innovation process, also with research institutions. Spatial proximity to the co-operation partners was irrelevant. Competence and excellence tended to be the decisive criteria. The second cluster of firms is characterised by medium outward orientation, and seems to rely more on in-house problem solving strategies. Such firms tend to have regular contacts with customers and suppliers. Linkages with research institutions and universities are less common. Geographical proximity to co-operation partners is less important. The third cluster relies almost entirely on in-house problem solving techniques. It includes less innovative firms with less complex products as well as highly specialised firms that operate in small market niches. Even though the latter are quite innovative, few have network activities in the competitive stage of the innovation process.

D The Location of Innovation Co-operation Partners

Most of the firms make use of various types of network relations in the innovation process. It can be seen from Fig. 2.4 that network relations are relevant at different spatial scales: the regional or metropolitan scale, the national scale, the European (EU and Central Eastern European) scale and the global scale. Some interesting observations can be made in this context.

Table 2.4 Network activities of manufacturing firms (1994-1996)

		Customer Networks		**Manufacturing Supplier Networks**		**Producer Service Supplier Networks**		**Producer Networks**		**Co-operation with Research Institutions**	
Pre-Competitive Stage			**c**		**c**		**c**		**c**		**c**
Information Exchange	**a**	199		135		165		66		61	
	b	64	(26.1 %)	45	(23.0 %)	63	(34.5 %)	27	(30.3 %)	25	(32.8 %)
Identification of New Ideas	**a**	190		122		148		64		57	
	b	57	(25.8 %)	39	(24.6 %)	57	(34.5 %)	25	(28.1 %)	20	(31.6 %)
Research and Development	**a**	179		118		148		49		56	
	b	55	(25.7 %)	37	(23.7 %)	56	(34.5 %)	20	(26.5 %)	22	(30.4 %)
Competitive Stage											
Prototype Development	**a**	175		108		96		37		47	
	b	53	(24.6 %)	34	(23.1 %)	36	(32.3 %)	16	(27.0 %)	20	(31.9 %)
Pilot Projects	**a**	167		97		101		28		47	
	b	51	(25.1 %)	30	(24.7 %)	41	(34.7 %)	12	(32.1 %)	20	(29.8 %)
Market Introduction	**a**	183		82		105		49		19	
	b	56	(26.2 %)	25	(25.6 %)	38	(34.3 %)	20	(22.4 %)	9	(31.6 %)

Notes: **a** denotes the number of such network activities of the manufacturing firms (with all regions);
b denotes the number of manufacturing firms with such network activities (with all regions);
c denotes the share of such network activities with a focus on the metropolitan region of Vienna.

- *First*, the *metropolitan scale* seems to play an important role in networking. Approximately one third of the co-operation projects with producer service providers and research institutes are regionally embedded. In the case of the other types of network relations, the proportion was about one quarter.
- *Second*, there is evidence that the European scale has become more important than the national scale in recent years as far as the location of innovation partners is concerned. This holds true not only for vertical interfirm relationships (i.e. customer networks, manufacturing and producer service supplier linkages), but also for the horizontal inter-firm relationships (i.e. co-operation with research institutes and producer network relations).
- *Third*, a small segment of network relationships can be observed at the *global scale*. Companies belonging to larger and foreign corporations are – due to their corporate links – generally more integrated not only into European, but also into global networks. In contrast with smaller firms they have only selective links in the metropolitan region, especially with the science & research sector.
- *Fourth*, it is important to note that most of the firms with networking activities maintain relationships at more than one of the spatial scales utilising different types of co-operation partner. It seems that regional, national, European and global networks are complementary rather than substitutive. There is some evidence that firms which have learned to co-operate at the regional scale find it easier to establish and maintain network relations at larger scales as well.
- *Finally*, it is worth mentioning that innovative firms (as defined above, i.e. where product innovations comprise more than 20 percent of turnover) establish not only more inter-firm links in the innovation process, but are also generally more embedded at all spatial levels.

2.5 The Producer Services Sector

Services not only provide linkages between different segments of production within a product chain, they may also co-ordinate and integrate the atomised production process. In several product markets, it is the services such as design, research, marketing, but also delivery and sometimes consumer credit that determine the competitiveness of manufacturing investment. As the length of production chains increases, so services tend to become responsible for a greater share of value added to products (Britton 1990). Many such producer services are deeply embodied in goods. The provision of these services may be carried out

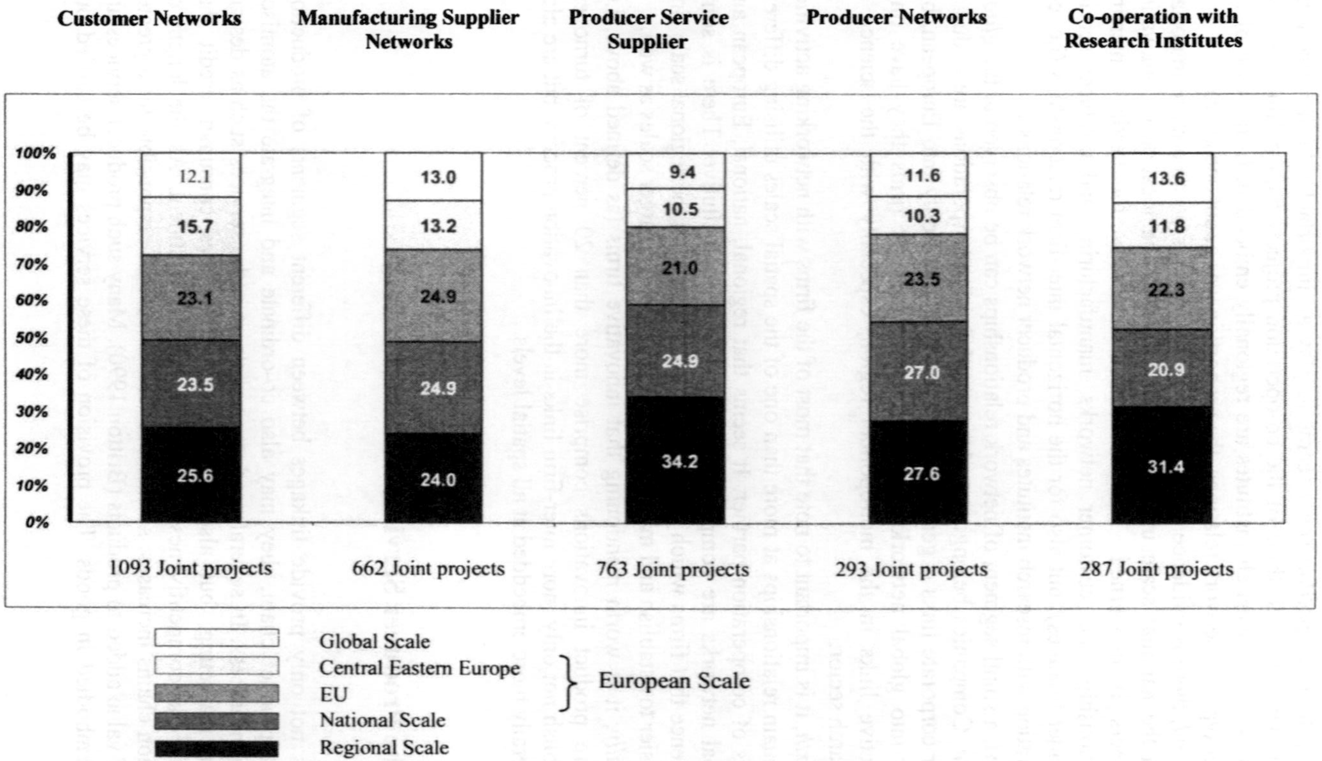

Fig. 2.4 Network activities by location of the innovation partners (1994-1996)

within the producing firm itself or may be externalised, i.e. put out to independent, specialised producer service providers. The boundary between the internalisation and externalisation of producer service functions changes continuously. Service functions may be externalised or, alternatively, specialised producer service providers may be taken over or they may operate on a subcontracting basis (see Dicken 1998 for more details).

The purpose of this section is to shed some fresh light on the sector of producer services within the Vienna metropolitan innovation system in general and to characterise the nature and extent of its innovation support activities for the manufacturing sector in particular.

A The Survey

There is a wide diversity of producer service providers in the metropolitan region of Vienna. This presents a major problem when attempting to identify and classify these producer service firms that act as innovation support units, and to measure the nature and extent of their innovation support activities. The survey elicited information on the size of producer service providers as well as the general nature and extent of innovation support or assistance offered to local manufacturing units, the structure of the customer/client base, the geographical location of the clients, and the nature and extent of innovation activities undertaken to improve their own service products and/or to offer new services.

Using information provided by the Austrian Central Statistical Office, we identified a population of 648 innovation support units within the business enterprise sector. These provide services such as computer software (hardware consulting, software consulting and supply, data processing, and database activities – NACE 72.1-72.4); technical consulting (engineering and related technical consulting, technical testing and analysis – NACE 74.2-74.3), business consulting (business and management consulting – NACE 74.14); and market research and advertising (market research, advertising, exhibitions and congress organisers – NACE 74.4, 74.8, and 74.13). Of these producer service providers, 190 returned the completed questionnaire, representing a response rate of 29.3 percent.

Table 2.5 presents the responses broken down into four producer service categories and four size classes, the latter defined in terms of employment. The sectoral distribution of sample firms roughly follows the patterns of the whole population, although the smallest producer service category, with less than 20 employees, tends to be underrepresented. The majority of producer service providers in the metropolitan region are very small: 73.1 percent have fewer than 20 employees. Most of the firms in the sample have been in business since 1970 (81.8 percent of those with a known starting year), and are located in the core area of the metropolitan region, i.e. the city of Vienna (86 percent). With respect to organisational status, 135 firms (71 percent) are independent, others are parts of larger organisations.

Table 2.5 Response patterns and response rate of responding producer service providers

	Total Number Registered Firms 1991		Number of Responding Firms 1997		Response Rate [a]
	no.	%	no.	%	
Producer Service Categories					
Computer Software	105	16.20	27	14.59	25.71
Technical Consulting	344	53.09	98	52.97	28.49
Business Consulting	104	16.05	28	15.14	26.92
Market Research and Advertising	95	14.66	32	17.30	33.68
Employment Size					
≤ 19	530	82.43	136	73.12	25.66
20-49	81	12.60	31	16.67	38.27
50-199	28	4.35	15	8.06	53.57
≥ 200	4	0.62	4	2.15	100.00

Note: a number of responding producer service firms divided by total number of firms multiplied by 100.

B Forms of Innovation Support or Assistance Offered to the Manufacturing Sector

There is wide agreement that the growing importance of the producer services sector is due primarily to the continuously intensifying demand for service inputs to production (Martinelli 1991). This reflects some fundamental changes occurring in modern economies, such as the increase in the organisational scale of production, the growing technical complexity of both the production process and the legal and financial regulation, and the rising uncertainty of the commercial and technological environment (Fischer 1990, Traxler et al. 1991). The increasingly complex nature of product and process innovations in manufacturing has played an important role in the recent expansion of the producer services sector. Substantially decreasing product life cycles complemented by pressure from the consumer side to intensify product differentiation tend to intensify the research and development activities, also design, advertising and marketing aspects of the production of goods. The increasing pace of technological change forces firms to seek specialised help or to develop specialised capacities in fields such as information processing, industrial engineering or process design (Bailly and Coffey 1991).

The complexity of industrial organisation, affecting particularly the suppliers of inputs, poses great challenges to smaller manufacturing firms, which often struggle to maintain competitiveness as innovation increases in importance. The knowledge needed to compete comes often from suppliers. Despite undertaking little or no R&D, many smaller firms are able to be innovative by relying heavily on innovation support units such as producer service providers. To a considerable extent, the level of competitiveness and innovative capacity depends on the degree to which firms are tied to local networks of suppliers. Local purchasing of producer services is important, because these local linkages foster innovation.

Tables 2.6 and 2.7 provide some empirical evidence on the type of innovation support for manufacturing clients from the point of view of producer service firms. They highlight the fact that:

- assistance is offered to manufacturing firms for product and process innovation activities in both the pre-competitive and the competitive stages of the innovation process. Technical consulting is widespread, but all other forms of producer services were also found. Computer software firms tend to be more often involved in process innovation issues, while market research and advertising firms were involved in product innovation.
- The size breakdown of producer service firms reflects the overall pattern of the producer service sector in the metropolitan region. Of firms providing innovation support, only a very low share had over 50 employees.
- Innovation support is often provided in co-operation with other service providers. Overall, there is only limited involvement with research institutes/ universities. University research groups were considered valuable both as an extension of the research capacity and as a kind of 'antenna' for accessing novel scientific knowledge. University research groups tend to have

worldwide contacts and easier access to emerging ideas, especially from abroad, and thus complement local producer service providers.

Table 2.6 Innovation support for manufacturing clients (1994-1996) by type of innovation and collaboration

	Innovation Support for				Innovation Support Provided in Co-operation with					
	Product Innovation		Process Innovation		Other Producer Service Firms		Public Research Institutes		None	
	a	b	a	b	c	d	c	d	c	d
Producer Service Provider Categories										
Computer Software	12	46	16	62	8	50	5	31	4	25
Technical Consultancy	30	32	25	27	24	57	17	40	15	35
Business Consultancy	11	42	15	58	9	50	4	22	9	50
Market Research and Advertising	17	61	10	36	16	89	5	28	2	11
Employment Size of Producer Service Providers										
≤ 4	13	37	13	37	8	42	6	32	8	43
5-9	24	40	20	33	18	62	8	28	11	39
10-19	12	35	15	44	11	69	2	13	7	46
20-49	15	54	14	50	14	70	7	35	4	20
≥ 50	7	39	5	28	4	40	7	70	1	10
N	106		106		190		190		190	

Notes: **a** denotes the number of producer service providers who supported manufacturing firms in their product or process innovation (1994-1996);
b denotes the number of producer service firms with such network activities (with all regions);
c denotes the share of such network activities with a focus on the Vienna metropolitan region;
d denotes percentage of such producer service providers out of total in the corresponding row category supporting manufacturing innovation.

C Location of Manufacturing Clients, Co-operation Partners and Innovation-Relevant Sources

In most countries producer services are concentrated in the largest urban areas. This reflects the costs of delivering these services to markets, as well as the need to locate near pools of highly skilled labour. In the Austrian case, producer service providers are overwhelmingly located in the metropolitan region of Vienna.

Table 2.7 Innovation support for manufacturing clients (1994-1996) in different stages of the innovation process

	Pre-Competitive Stage								Competitive Stage			
	Information Exchange		Identification of New Ideas		Research and Development		Prototype Development		Pilot Projects		Market Introduction	
	a	b	a	b	a	b	a	b	a	b	a	b
Producer Service Provider Categories												
Computer Software	12	71	15	88	16	94	15	88	12	71	7	41
Technical Consultancy	20	44	28	62	32	71	27	60	14	31	4	9
Business Consultancy	14	74	15	79	14	74	10	53	4	21	9	47
Market Research and Advertising	16	73	17	77	19	86	10	45	14	64	22	100
Employment Size of Producer Service Providers												
≤ 4	7	64	9	82	8	73	4	36	6	55	5	45
5-9	19	63	21	70	22	73	21	70	13	43	11	37
10-19	17	57	22	73	23	77	16	53	12	40	13	43
20-49	13	62	15	71	19	90	11	52	9	43	7	33
≥ 50	6	55	8	73	8	73	8	73	4	36	6	55

Notes: **a** denotes the number of firms indicating innovation support in industry;
b denotes the percentage of innovation supporting firms of all responding firms in the corresponding row category.

n = 103

Table 2.8 Location of manufacturing clients and co-operation partners (1994-1996)

	Manufacturing Clients (n=151)		Producer Service Co-operation Partners (n=184)		Public Research Institute Partners (n=188)	
	a	**b**	**a**	**b**	**a**	**b**
Metropolitan Scale	653	52	376	53	37	56
National Scale	300	24	219	30	13	20
EU	157	13	80	11	13	20
Central Eastern Europe	97	8	24	3	1	2
Global Scale	36	3	22	3	1	2
Total	1,243	100	721	100	65	100

Notes: **a** denotes the number of manufacturing clients, producer service co-operation partners and public research institute partners, respectively;
b denotes the corresponding column percentage.

Table 2.8 provides evidence that inter-firm interaction – via input-output linkages as well as information exchange – is largely a metropolitan phenomenon, especially for smaller producer service providers. If firms are in close and frequent communication with their manufacturing clients and their co-operation partners (i.e. other producer service firms and public research institute partners), the risk of an innovation failing to meet a client's needs are minimised. The web of local information flows, co-operation linkages and subcontracting links is at the heart of the agglomeration advantages characteristic of the metropolitan region of Vienna. Network relations with other producer service firms seems to be a popular form of inter-firm co-operation, while producer service and research relationships are not very frequent.

Keeping up with state-of-the-art developments is critical not only for manufacturing firms but also for producer service providers. Fig. 2.5 provides evidence on four major sources of innovation-relevant information: industrial clients, direct competitors, other producer service firms and universities/research institutes. The most important sources used by producer firms are research institutes and, rather less frequently, other producer service firms. It is interesting to note that the information needs of producer service firms are not markedly different from those of manufacturers. However, producer service providers rely much more on information sources located in the metropolitan region than manufacturing firms.

Producer service firms – because many of them rely on local customers and co-operation partners – are more able to benefit from nonspecific information traded at informal gatherings. The data clearly shows that spatial proximity seems to be an important criterion for utilising external sources of innovation-relevant information, independent of the sources.

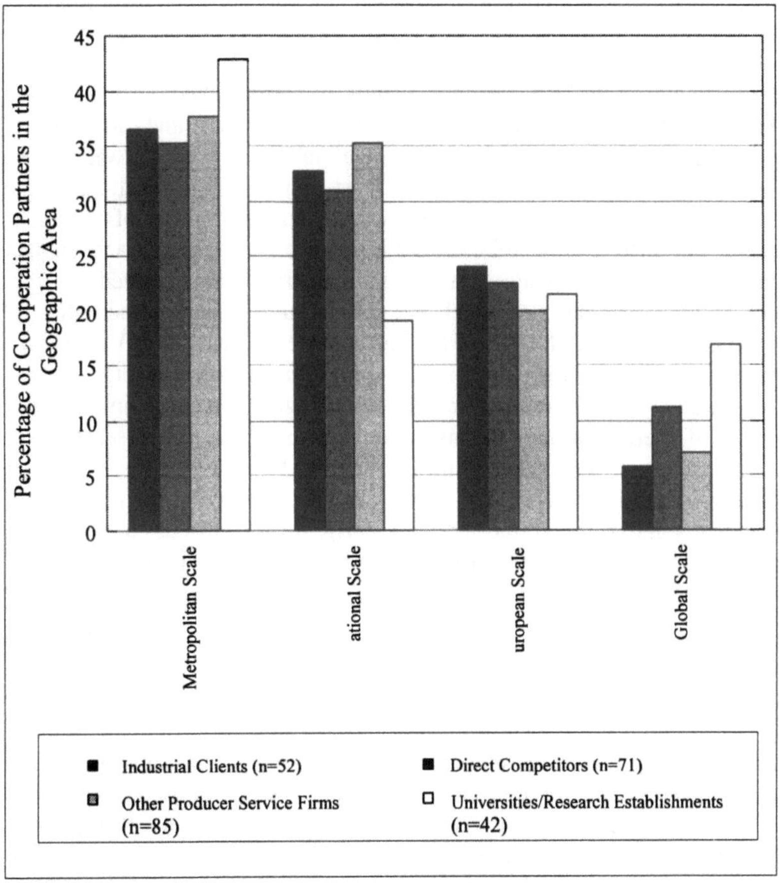

Fig. 2.5 Sources of innovation relevant information by spatial scale (1994-1996)

2.6 The Science & Research Sector

Science & research activities figure prominently in Vienna's past. Its first university, the University of Vienna, was founded in 1365, followed by the Academy of Fine Arts in 1692, and the University of Veterinary Medicine in 1765. The period of greatest growth, however, was the 19th century when there was a quantitative expansion and qualitative improvement in further education. This began with the opening of the Vienna University of Technology, founded in 1815, was then followed by the University of Applied Arts, founded in 1868, the University of Agricultural Sciences, founded in 1872, and a Business School in 1898. This latter was mainly an initiative from the chamber of commerce and individuals from the world of commerce and industry. The Business School

(Hochschule für Welthandel) later developed into a university-level institution and education at this school became a routine means of entry into business management and government in Austria.

By the beginning of the 20th century Vienna had established a sophisticated system for education and scientific, technical, commercial and art matters, reaching from elementary school to the doctoral level. There were close connections between the different levels in most areas of specialisation, as the teachers for the schools at a given level were usually educated at one of the higher levels. There was also a flow of knowledge between universities and subject fields. Links between the education system and industrial firms operated not only through the supply of trained personnel, but also through consulting activities by professors in engineering and areas of applied science. At that time, Vienna was one of the leading centres of science in the world. Many scholars went to Vienna to learn the state-of-the-art. Viennese institutions of higher education, in particular the University of Vienna, became the institutional focus of scientific research.

Due to lack of space, we cannot bring the historical account of the science & research sector up to the present time. Instead we briefly describe the main actors and then present some details of the survey. Finally, we shall discuss the results in order to shed some light on the structure of this important component of the metropolitan innovation system and its linkages with the business enterprise sector.

A The Major Actors and Their Roles

The metropolitan region possesses a diverse range of research institutions that are committed to different technology transfer channels. The public funded science & research sector includes two major groups: universities and non-university based research establishments. Universities that combine their educational function with the advancement of scientific knowledge play a double role in the innovation system. *First*, they train and educate scientists and other technical personnel, expanding the basis of scientific and technical information. *Second*, they transfer their talent and information to the business enterprise sector, thereby fostering the commercialisation and diffusion of innovation. This co-operation between the educational and industrial sectors is not a new phenomenon, but the need for this interaction has become greater today. Universities, industry and the state have started to develop programmes to address not only the preparation of students, but also the new and expanding demands for increased R&D and technology transfer, and the need to provide the regional workforce with technical skills. As budgetary pressures have slowed down federal R&D funding in recent years, the university sector has become more active in soliciting research funds from industry.

The University of Vienna has a tradition of research in natural and medical science reaching back to the Middle Ages. This university is known mostly for teaching and basic research without aiming at any commercialisation. Its main transfer channel seems to be the publication of research results. The share of research grants from industry remains low. In contrast, the Vienna University of Technology has a long tradition of industry-related research. Similar to the US, technical universities have been founded in the German speaking countries since the 19th century to enforce inventions and technical applications of scientific findings, but have subsequently focused on basic research. They receive external research funds from industry, though this represents generally less than 10 percent of all their research spending. The Vienna University of Economics and Business Administration (the former Hochschule für Welthandel), the University of Agricultural Sciences and the University of Veterinary Medicine may be positioned somewhere between these two extremes.

In the vast majority of cases, teaching and research are carried on by the smallest units in the university the institutes/departments. Knowledge transfer from academic research to private firms flows through many channels. New knowledge generated in universities is transferred to the business world through the publication of research papers, R&D contracts or co-operative R&D with private companies. Informal contacts and the hiring of researchers is more often used to get access to uncodified and person-embodied forms of knowledge. The Vienna University of Economics and Business Administration is also known for its postgraduate programmes, but it is usually suggested that university research falls short of the full impact it could or should have.

In the metropolitan region of Vienna – like in other regions and countries across Europe – inefficiencies in the transfer of technology and lack of orientation towards the commercialisation of scientific results are held to be the predominant reasons for ineffective public research, rather than the lack of quality of research in terms of scientific performance. It can be argued that there are no incentives for scientists to support commercialisation when they are solely evaluated in terms of scientific performance. There is a broad consensus that new structures are needed to give universities more responsibilities and at the same time further increase the incentives to act in a more efficient and competitive way in the future.

As tradition-bound universities offered limited potential for ambitious research policy, additional research establishments were founded by governments to fill gaps in the wide spectrum of technology transfer to private business. The history of public funded research centres in the metropolitan region goes back to the last century. With only a few exceptions, most notably the research institutes, research centres and scientific commissions of the Austrian Academy of Sciences, the justification for their establishment was either to maintain the international competitiveness of private companies or to close a technological and/or scientific gap between Austria and other nations. Among the most prominent post World War II examples are the Austrian Research Centre Seibersdorf, set up in 1956 as the Society for the Study of Nuclear Energy, the Arsenal Experimentation and Research Institute with its vehicle research and testing facilities, founded in 1950, and the Federal Environment Agency set up in 1985.

Founded under a decree dated 14 May 1847 as the 'Imperial Academy of Sciences in Vienna', the Austrian Academy of Sciences is one of the key players in the science & research sector. Its present legal status was laid down in the Federal Act of 1921 and requires the Academy to further science in every aspect. In doing so, the Academy has the right to the protection and assistance of the federal authorities. It carries out its activities on the basis of a statute. Within the framework of this statute, and preserving the tradition of a learned society, the Austrian Academy of Sciences has grown, particularly during the last decades into Vienna's and Austria's most important non-university research institution. At present its research work is organised in the form of 20 institutes, four research centres and 30 scientific commissions, generally located in Vienna. Including the national and international programmes co-ordinated by the Academy, the number of employees is over 500. Though its activities are diverse, the main concern is to concentrate on fundamental research and on those fields that are insufficiently represented in the universities.

B The Survey and Some Characteristics of the Research Units Surveyed

The feasibility of a postal questionnaire to survey research institutions depends on the ability to design a schedule that is sufficiently general to be applicable to a wide range of organisations. For some, e.g. universities or other large research organisations, it did not make sense analytically to regard the whole organisation as a single reporting unit. In such cases, it was necessary to define subunits such as departments or institutes as the relevant object of the survey. Considerable preparatory work was required to identify the population of relevant innovation support research units, i.e. research organisations with appropriate departments or units to target. Based on information provided by the Austrian Statistical Office, we identified a population of 650 such research units. The overwhelming majority of these are located in the City of Vienna, the core of the metropolitan region. Nearly half of these belong to the university sector (about 45 percent), the rest were part of public non-university organisations (about 40 percent) or business-related research centres (about 15 percent). Of these research units, 290 returned the completed questionnaire, representing a response rate of 44.6 percent.

Table 2.9 presents the responses broken down into three science & research segments, i.e., the university sector, public non-university organisations, and business-related research centres. Four spatial categories were defined [the 1st District, 2nd to 19th Districts and 10th to 23rd Districts of the City of Vienna, and the urban fringe]. The sample can be seen to broadly reflect the overall structure of the total population.

The survey served to elicit information on the general nature of innovation services offered by local units of the science & research sector. The questionnaire was targeted to research departments/units within the following major science fields: architecture, construction, surveying; biology, chemistry and medicine; mathematics, informatics and physics; electrotechnology and mechanical engineering; social sciences including economics and geosciences. Of the research units surveyed, 27.6 percent employed only three or less scientific staff members including professors/leading staff, and 36.0 percent more than ten scientific staff members. On average every third scientific staff member is funded by third parties, i.e. funds from business, public funds from the region, such as from the City Government, national funds, such as the Austrian Science Fund and the fund provided by the Austrian National Bank, funds of international organisations (e.g. the EU) and other funds.

Table 2.9 Response patterns and response rate of responding research units

	Total Number of Registered Research Units 1997		Number of Responding Research Units 1997		Response Rate[a]
	no.	%	no.	%	
University Institutes/Departments	290	44.61	132	45.52	45.52
Public Non-University Research Organisations and Business-Related Research Centres	259	39.85	117	40.34	45.17
Industry-Related Research Centres	101	15.54	41	14.14	40.59
Geographical Location					
Vienna - District 1	95	14.62	43	14.83	45.26
Vienna - Districts 2-9	372	57.22	169	58.28	45.43
Vienna - Districts 10-23	153	23.54	63	21.72	41.18
Vienna Surroundings	30	4 .62	15	5.17	50.00
Total	650	100.00	290	100.00	44.62

Note: a number of responding units divided by total number of units multiplied by 100.

Table 2.10 Third-party funded or contract research, measured in terms of scientific staff (1994-1996)

	Research Units					
	Architect. Construct., Surveying	**Biology, Chemistry, Medicine**	**Maths, Informatics, Physics**	**Electro-technology, Mechanical Engineering**	**Economics, Social and Geosciences**	**Total**
***Third-party Funded Scientific Staff* (% total)**						
Business Funds	14	25	19	40	7	20
Public Funds from Region	20	13	16	5	13	13
National Funds	45	42	45	42	53	45
International Funds	7	11	15	12	13	12
Other	14	9	5	1	14	10
Total	100	100	100	100	100	100

Table 2.10 provides evidence on the relative importance of these types of funds across the various scientific fields. Clearly, the percentages vary a great deal among research units across the scientific fields as well as between the university sector and the non-university sector. With respect to the university and non-university sectors, the differences are clear. The non-university sector is more dependent on the support of third parties for research than their counterparts in the university sector. But universities produce a double product, in the sense that research and teaching are produced together. Thus, it is far from easy to fully solve the problem of the imputation of costs in these cases. Table 2.11 shows how the total time budget is spread over the following activities: basic research, applied research, teaching activity, transfer tasks (e.g. presentation of research results) and other activities.

C Innovation Support or Assistance offered to the Manufacturing Sector

The fact that the linear innovation model has fallen into disuse is no doubt attributable as much to the improved understanding of reality as to a change in reality itself. The change is in any event a 'real' one and makes it even more urgent to grasp all the policy implications of the non-linearity of the innovation process. Simply stated, government must now become more involved in the marketing of scientific know-how, and firms need to participate more directly in basic research.

The combination of these two trends is increasing the importance of pre-competitive research and development, especially when it concerns so-called generic technologies. To increase the stock of public scientific knowledge which is accessible to all the actors in the innovation process, it is necessary to finance the activities of research establishments dedicated to long term basic research, such as universities and non-university public research institutions like the Austrian Academy of Sciences.

Table 2.11 Total time budget of scientific staff spread over research, teaching and other activities (in %) (1994-1996)

	University Sector	Public Non-University and Private Non-Profit Organisations	Business-Related Research Centres
Basic Research	24	26	12
Applied Research	24	44	63
Teaching	32	7	5
Transfer Tasks	6	10	8
Other	14	13	12

n=276

The metropolitan region has a strong presence of high quality science & research establishments. But the research base is not integrated into the innovation system as well as it could be. In fact the business world and academia are very distinct in nature. While in firms, problem-solving in a rapidly changing environment is the rule, and flexibility rather than specialisation is highly valued, academic research is generally carried out in a controlled world where continuity and specialisation are important. Academic scholars are rewarded through publishing original ideas or novel findings, while industry has little or no interest in novel ideas related to basic research and often seeks to delay or even prevent publication.

The increasing complexity of the scientific base for industrial activities increases the innovator's dependence on knowledge, and technology incorporating that knowledge, that he/she does not possess in his/her own right. Thus, increasing attention is being given to those disciplines and scientific fields that can bridge the gap between the type of knowledge produced by basic science and the type of knowledge needed by manufacturing firms in day-to-day innovation activities. Industrial R&D laboratories and industry-specific research associations are geared to bridge the gap from the industry side. However, the bridging process has also to be organised from the side of universities and public non-university research establishments. Under growing budget constraints these establishments have increasingly grasped the need for relating with industry.

Table 2.12 Innovation support to manufacturing clients (1994-1996) in different stages of the innovation process

		Universities	Public Non-University Establishments	Business-Related Research Centres
Pre-Competitive Stage				
Information Exchange	**a**	17	5	6
	b	(12.9%)	(4.3%)	(14.6%)
Identification of New Ideas	**a**	19	8	6
	b	(14.4%)	(6.8%)	(14.6%)
Research and Development	**a**	24	6	5
	b	(18.2%)	(5.1%)	(12.2%)
Competitive Stage				
Prototype Development	**a**	12	4	10
	b	(9.1%)	(3.4%)	(24.4%)
Pilot Projects	**a**	14	4	7
	b	(10.6%)	(3.4%)	(17.1%)
Market Introduction	**a**	5	1	4
	b	(3.8%)	(0.9%)	(9.8%)

Notes: **a** denotes the number of research units with innovation support to manufacturing clients; **b** denotes the share of research units with innovation support in the corresponding column category. n=169

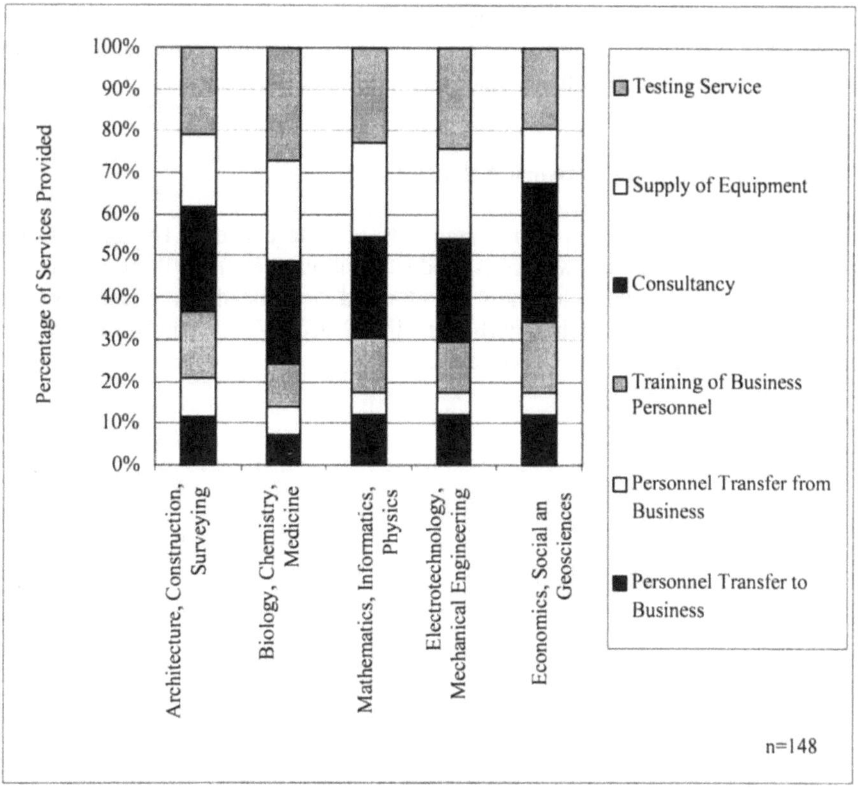

Fig. 2.6 Breakdown of services or forms of assistance offered to manufacturing firms by science fields (1994-1996)

Co-operative research activities may provide industry with early access to the results of research and improved training for scientific and engineering personnel. But there are no signs up to now that the science & research sector already encompasses the mission of economic development in addition to research and teaching, in other words, the transfer of scientific and technical knowledge to firms that need the knowledge.

Table 2.12 illustrates that the science & research sector provides innovation support to manufacturing firms only at a rather modest level of intensity. There are weaker links between the partners and a focus on the earlier stages of the innovation process. Research establishments appear to have the most efficiently managed linkages with larger firms operating in high tech fields, both in the metropolitan region and beyond. These firms tend to view university research groups as valuable both as an extension of their own research capacity and as some sort of antenna. Firms recognise that because senior university scholars belong to well established scientific networks, and are likely to have worldwide contacts, they may have easier access to new ideas, especially from abroad. Technical consultancy (i.e. disembodied information) and test services are widespread, but other forms of support, such as training and personnel transfer to and from the research establishments, were also found (see Fig. 2.6).

D The Geography of Network Activities

Until recently, the location of research was of little concern. The relationship between the site where knowledge is produced and its eventual utilisation was not seen to be tightly linked. This view has changed dramatically in recent years, as has the notion that high tech conurbations like Silicon Valley and Route 128 are unique instances that cannot be replicated.

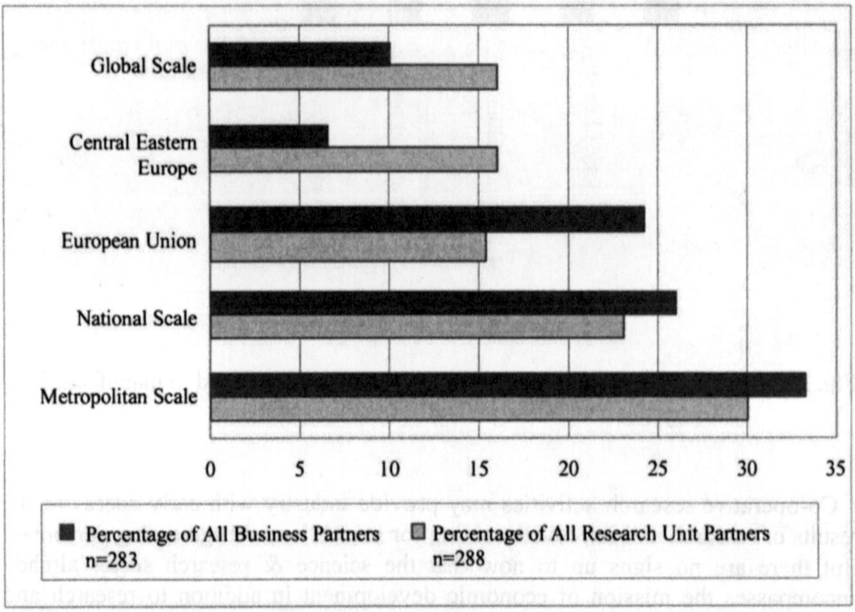

Fig. 2.7 Location of co-operation partners of research units with intensive and very intensive research links (1994-1996)

Successful technology transfer implies not just the receipt of a set of blue prints, but the development of the ability to apply and explicit the commercial possibilities of a new technology. Most technologies consist of codified and tacit components. The transfer of a technology requires access to the tacit components as well as to those codified in a blue print or data package. The importance of this tacit component means that technology transfer often requires the transfer of people as well as license agreements.

Fig. 2.7 provides evidence on the geographical extent of innovation support to manufacturing firms. As expected, research establishments serve mainly clients located in the metropolitan region (33 percent), Austria (26 percent) and the EU (24 percent), although there are also links to Central Eastern Europe (7 percent) and to higher spatial scales (10 percent). In particular, university-industry interactions seem to be facilitated by spatial proximity. Due to the more specialised nature of co-operation within the science & research sector, research partners appear to be somewhat less important at the metropolitan scale (30 percent), the national scale (23 percent) and especially the EU-level (15 percent), while partners at the global scale (16 percent) and in Central Eastern Europe (16 percent) tend to become more important.

2.7 Concluding Remarks

In this chapter we have looked in some detail at the four key building blocks of the mtropolitan innovation system of Vienna: the institutional sector, the manufacturing sector, the producer services sector, and the science & research sector.

The central actors in the innovation system are the manufacturing firms and their R&D laboratories. The majority of these firms are small – about two thirds have less than 100 employees and many of these have been in business since 1970. In terms of organisational status about half of them are independent, the other half operates within a wider parent company group, typically as a main plant rather than as a branch plant. The sectoral distribution indicates a predominance of innovative firms in electrical and optical equipment, machinery and transport, and chemicals, plastics and rubber products. These sectors account for more than 60 percent of all innovative firms. Firms were classified as innovative if product innovations introduced during the period 1994-1996 comprised more than 20 percent of the firm's annual turnover. Some firms have set up R&D laboratories on site, but these laboratories tend to be biased towards applied research and development. Small firms that operate in niche markets cannot afford large R&D laboratories. Even larger firms were mostly risk-averse and short-term oriented, and would not support the large investments necessary because of indivisibility and high uncertainty. A significant amount of R&D is performed by affiliates of foreign companies. Basic research that contributes to the world stock of knowledge is largely concentrated in the science & research sector or in the research laboratories of foreign multinationals, often outside the region. Smaller firms are unlikely to undertake research with external applicability or geographically distant commercial benefits.

Manufacturing firms in the Vienna metropolitan innovation system tend to be good at improving their existing product lines and building incremental innovations into their products, sometimes tied to changes in production technology and new organisational practices. But, in general, these innovation activities are based on product and/or process innovations of the kind that builds on knowledge embedded in the labour force, rather than on radical innovations in new product lines. Such innovation strategies are favoured by the institutional set-up. The Austrian partnership model limits the capacity of firms to employ workers in tasks that cut across skill categories or shift workers from one task to another, and, moreover, discourages lay-offs.

In the context of new challenges that have arisen with the opening up of Eastern Europe and Austria's membership of the EU, flexibility will become more important in future than it has been in the past. This will call for the social partnership - one of the distinctive features of the Viennese innovation system - to adapt to new challenges by introducing more flexibility at the micro level, while continuing to rely on the strengths of the existing co-operative, associational and consensus-oriented institutional framework at an aggregate level.

Even if innovation is still very much an internal process for the firms in the metropolitan region, there is evidence of networking activities of various kinds. Some of the results obtained are worth underlining. *First*, co-operation in the pre-competitive stage of the innovation process is generally more common than in the competitive stage, independent of the type of networking. External information tends to be particularly relevant during the early stages of the innovation process, when perception of problems and evaluations of technological possibilities take place.

Second, customer networks, but also user-supplier networks (i.e. producer service and manufacturing supplier networks) are much more frequent than horizontal forms of co-operation, such as producer networks and research institution-industry linkages. Customer networks represent the most frequent form of inter-firm co-operation. Manufacturing suppliers and producer service providers have strong incentives to establish close relationships with user firms and even monitor some aspects of their activity. Knowledge produced as a result of learning-by-doing can only be transformed into new products if the producers have direct contact with users. In turn, user firms will generally need information about new products or components. This may not only mean awareness, but also quite specific inside information about how new, user-value characteristics relate to their specific needs. User-manufacturing supplier linkages frequently involve foreign organisations as users or suppliers. Many are also involved in the commercialisation of technology developed in the science & research sector of the region.

Third, consumer and user-supplier relationships basically involve two types of interaction. One is interdependent, functioning as a co-operative mode relying on tacit performance agreements, trust and reciprocal adjustment. The second is more of a contractual, competitive or arm's length mode, where interfirm trust and familiarity may be very limited. Both types of transactions appear to coexist in the metropolitan region.

Fourth, the firms with networking activities tend to maintain linkages at more than one of the spatial scales, utilising different types of co-operation partner. Regional, national, European and global networks are complementary rather than substitutes. The metropolitan region, thus, may be seen as an open system with important national and European linkages at the micro-level of manufacturing firms.

The metropolitan region has a well developed organisational infrastructure of mediating organisations for technology and vocational training and a strong presence of high quality research and educational organisations. But the research base is not integrated into the innovation system as well as it could be. The industry-university linkages are one of the weaker elements of the metropolitan innovation system. In fact, the business world and academia are very distinct in nature. While in firms problem-solving in a rapidly changing environment is the rule, and value is given to flexibility rather than specialisation, university research tends to be carried out in a controlled world where continuity and specialisation are important. Academic scholars are rewarded through publishing original ideas or novel findings, while industry has little or no interest in novel ideas related to basic research and often seeks to delay or prevent publication (Cooke, Boekholt and Tödtling 2000). Evidently, there is a need to bring the centres of knowledge production closer to those who might use this knowledge for commercial purposes, without compromising the essential scholastic and critical function of research based universities and research establishments.

3 The Barcelona Metropolitan System of Innovation

This chapter provides statistical evidence on the innovation system of Barcelona. The case study region comprises the communities of Barcelones, Maresme, Valles Oriental, Valles Occidental and Baix Llobregat which, from a functional point of view, form the metropolitan region of Barcelona (see Fig. 3.1). Two thirds of the Catalan regional product comes from the metropolitan region of Barcelona, which at the same time concentrates the largest number of research institutes in Catalonia. After an overview of the economic performance and structure of the Catalan capital (Section 3.1), the Spanish and Catalan innovation system is briefly characterised in Section 3.2. Special attention is paid to the development of regional institutions. This development is a consequence of the decentralisation process introduced by the 1978 Constitution, as well as the increase in international competition following Spain's entry into the European Community. As a result of the innovation-averse economic policy of the Franco era, the major challenge has been the transformation of the innovation system. The innovative capabilities in the business and science sector are less developed and not as well linked with each other as they are in more advanced regional economies.

Surveys of the main actors of the metropolitan innovation system, the manufacturing firms (Section 3.3), the producer service firms (Section 3.4) and the research institutions (Section 3.5), illustrate that drastic changes have taken place with respect to innovation and network activities. The catching-up process is expressed in impressive growth rates for the metropolitan economy and is accompanied by a greater innovative orientation of the actors surveyed. But despite this positive development trend, the metropolitan region of Barcelona is lagging behind the European core regions.

Section 3.6 highlights the question: to what extent is the propensity to co-operate with external partners in innovation processes influenced by firm specific factors? External co-operation is becoming more and more important due to the increased complexity of the innovation process. As discussed in Chapter 1, this complexity leads to a higher degree of interaction between the various actors involved, including the customers, manufacturing suppliers, producer service providers, and research institutions. On the basis of the empirical analyses, the strengths and weaknesses of the Barcelona metropolitan innovation system are summarised.

3.1 Barcelona – the Dynamic Capital of Catalonia

The region of Catalonia, whose economic, political and social centre is the metropolitan region of Barcelona, is Spain's leading industrial region. In 1995 Catalonia achieved around 25 percent of the country's industrial net value added with 23 percent of the industrial employees (compared to Madrid's 11.4 percent and 10.8 percent respectively). It accounted for approximately 27 percent of Spanish exports and in 1993 achieved around 20 percent of Spain's gross domestic product (the same figures for Madrid were 9.7 and 16.5 percent). In terms of per capita gross domestic product, in 1996 Catalonia held the second position among Spanish regions, after Madrid (see Table 3.1).

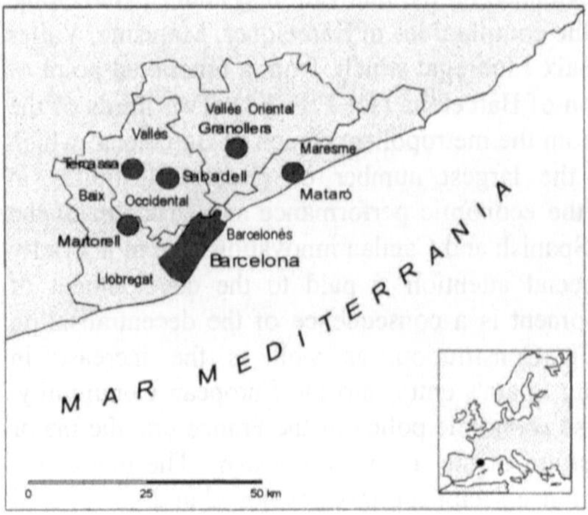

Fig. 3.1 The Barcelona metropolitan region

This state of development, although in line with the average when compared with values in the European Union (per capita GDP is 99 percent of the EU average) is considerably weaker than the other two regions examined in this volume: Stockholm (133 percent of the EU average) and Vienna (167 percent). Of the Spanish regions, Catalonia has one of the best economic performances, which has been around the EU average since the mid 1980s, but is distinctly weaker than the economic core regions of Europe (Eurostat 1999).

For Catalan firms, worldwide changes, especially the increased economic globalisation, present a very special challenge. Firms are mostly small family businesses and have for decades been directed towards the provision of local, regional or, at most, national markets. Until a few years ago they were protected from foreign competition by high duties and restrictive laws concerning capital invested abroad. It was only with Spain's entry into the European Community (EC) on January 1st 1986, that the economic policy, which until then had been aimed at autonomy, was abandoned. Above all, after the Single Market came into effect on January 1st, 1993, firms had to adapt to a new and extremely dynamic framework that was characterised by highly competitive conditions (Garcia 1989; Bienefeld 1995).

Table 3.1 Regional economic indicators of Catalonia

	GRP per Capita 1996 EU = 100	Sectoral Employment 1997 in % Agriculture	Industry	Services	Unemployment Rate in % 1997
Andalucia	57	12	22	66	32
Aragon	89	10	34	56	14
Asturias	74	11	30	59	21
Baleares	97	2	24	73	12
Canarias	74	8	18	74	21
Cantabria	77	11	30	59	21
Castilla-La Mancha	66	12	33	55	19
Castilla y Leon	76	14	28	58	20
Catalonia	99	3	38	58	17
Extremadura	55	16	25	59	30
Galicia	63	22	28	50	19
La Rioja	89	9	39	52	12
Madrid	101	1	27	72	18
Murcia	67	12	27	61	18
Navarra	98	9	40	51	10
Pais Vasco	92	3	37	60	19
Com. Valenciana	74	6	34	60	21
Spain	79	8	30	62	21
EU-15	100	5	29	65	11
Barcelona	70	3	36	61	21
Vienna	167	0	22	77	6
Stockholm	133	1	16	83	8

The consequences of Franco's economic policy still have an influence even today. Family firms, in particular, do not consider innovation or technological change a key to success. Spanish businesses therefore have to face productivity disadvantages and antiquated concepts of business management. Only very late have most adopted measures to increase their productivity and competitive ability. Many have not even tried and have been bought up by foreign businesses (Garcia 1989).

Despite the lack of competitiveness and internationalisation, a Spanish 'economic miracle' occurred after Spain became a member of the EC, and businesses established in Catalonia profited from this to an above-average extent. By the beginning of the nineties, Catalonia was able to achieve above-average growth rates in its gross regional product, compared with the EU, as a result of the internationalisation strategy of the Catalan government, which led to a sharp rise in direct foreign investments. However, this gratifying development is marred by a very high unemployment rate and by large sectoral differences (some sectors, above all heavy industry, are currently going through a serious structural crisis) (Caravaca and Sanchez Lechuga 1995). In spite of the catching-up process, Spanish businesses are still characterised by relatively low productivity and competitiveness in comparison with other OECD countries (Garcia 1989).

Table 3.2 Industrial structure in Catalonia (1995)

	Employees		Firms	
	in '000s	%	no.	%
Industry Sector				
Textiles & Clothing	91.8	18.2	6,579	22.0
Food Industry	72.9	14.4	4,487	15.0
Wood, Paper & Printing	62.6	12.4	5,705	19.1
Chemicals, Plastics & Rubber	87.6	17.4	2,326	7.8
Electrical & Optical Equipment	43.5	8.6	1,729	5.8
Basic Metals & Metal Products	62.3	12.3	5,799	19.4
Machinery & Transport	83.9	16.6	3,294	11.0
Total	504.6	100.0	29,919	100.0

The current economic structure is characterised by a relatively strong industrial sector (29 percent of the Catalan GRP in 1995) (see Table 3.2). Measured in terms of employees, the most important sectors are textiles (91,800 employees), chemicals, plastic and rubber (87,600 employees), machinery and transport (83,900 employees), food industry (72,900 employees) and the wood, paper and printing sector (62,600 employees). Machinery and transport, textiles and chemicals, plastics and rubber are sectors with strong export figures. On the other hand, the high-tech sector, including information and communication technology, as well as biotechnology, is only weakly represented.

Catalan industry is concentrated in Barcelona and the surrounding communities of Baix Llobregat, Valles Occidental, Valles Oriental and Maresme, which form our study region. Roughly 75 percent of Catalan industrial production comes from this metropolitan region which employs more than 80 percent of the industrial workers in Catalonia. Apart from this, there is one other relatively important industrial area (the petrochemicals industry) in the triangle formed by Tarragona, Reus and Valls south of Barcelona.

The Catalan service sector (including trade and tourism, transport and communication, finance, public administration and education) contributes 57 percent to the GRP of Catalonia (1995) and employs 60 percent of the workforce (1996). In producer services, there is a clear specialisation pattern towards business-oriented services, such as marketing and management consultancy. In 1992 nearly 70 percent of the registered producer service firms were engaged in business-oriented services. Technical consulting in the engineering sector and informatics is less important. The producer service firms are concentrated in and around Barcelona (see Table 3.3).

Table 3.3 Producer service firms in Catalonia (1992)

	Number of Firms		
	Catalonia	Barcelona	Total Catalonia %
Producer Service Sector			
Computer Software	421	305	72
Technical Consultancy	683	499	73
Business Consultancy	1,706	1,112	65
Market Research and Advertising	758	660	87
Total	3,568	2,576	72

The high level of dependence on foreign technology is a typical feature of Catalan industry. It is possible to identify three foreign investment waves which occurred during the 20th century. Shortly after the turn of the century there was investment from firms in the pharmaceutical industry, the food industry and electrical engineering. Well-known examples are Pirelli (1902), Siemens (1910) and Nestlé (1920). Franco's goal of self-sufficiency and Spain's political isolation after World War II were followed by the opening up of the country at the end of the fifties. Franco's aim was to develop an independent industrial base with the help of foreign investors as part of an import substitution development strategy. This led to foreign investment, especially in the chemical industry and the production of consumer goods and cars. The SEAT factory (initially a subsidiary of the Italian car manufacturer Fiat, and today a member of the Volkswagen Group) was established in Barcelona at this time. A third investment wave began after Spain joined the EC in the mid eighties and focused on consumer electronics, paper, chemicals, motorcycles, cars and automobile suppliers. NISSAN, VW, Benckiser and Honda were among the firms concerned.

A characteristic of the majority of these investments was that they were take-overs of existing businesses, which meant that practically no completely new production sites were set up in that period (Bienefeld 1995; Molero and Buesa 1996a). More recently, the importance of Catalonia as a target region for foreign investors has diminished. Whereas in 1991, 43 percent of Spain's foreign investment was in Catalonia, mostly in the area of Barcelona, this very high percentage was reduced to 25 percent by 1995 as a result of increasing competition within Spain, but also with Portugal and Eastern Europe.

There is an extremely strong dependence on foreign technology in high and medium tech sectors such as electric tools, cars and trucks (see Table 3.4). Due to growing internationalisation and massive foreign direct investments, the chemical industry too is increasingly dependant on foreign technology. One reason for this is the presence of multinationals. However, despite the technology-intensive products manufactured in Catalonia and the innovative production processes introduced there, the multinationals remain rather isolated. Because of their close functional links with the parent companies, these businesses contribute very little to the development of a regional network of technology providers and users, including the small specialised firms. With the exception of these large foreign multinationals, the processing business in Catalonia is characterised by small and medium-sized firms, which hardly carry out any R&D activities.

Table 3.4 Technological dependence of the Catalan economy (1998)

	Imports	Exports [in million Euros]	Foreign Trade Balance
Technological Level of Imported and Exported Goods [a]			
High	7.00	4.65	-2.35
Medium	14.76	12.60	-2.32
Low	10.75	8.43	-2.33
Total	32.51	25.68	-6.83

Note: **a** The OECD classification of goods is based on indicators of technology intensity. Goods like aircraft, office and computing equipment, drugs and medicine, radio, TV and communication equipment are classified as high technology goods. Motor vehicles, electrical machines, rubber and plastic products, ships, metal products are classified as medium technology goods. Paper, paper and printing products, textiles, leather products, food, beverages and tobacco, wood products and furniture are classified as low technology goods. For further details on the definition of technological levels of goods, see Hatzichronoglou (1997).

Between 1990 and 1995, R&D expenditure in Catalonia was 0.91 percent of the gross regional product, and thus slightly higher than the average value for Spain (0.87 percent of the gross domestic product). When compared with the national average in Austria (1.52 percent in 1996) and Sweden (3.60 percent in 1995), and even compared with the EU average (1.84 percent in 1995), Spain is clearly lagging behind. The metropolitan regions of Madrid and Barcelona are the leading R&D locations in Spain. Using indicators like R&D expenditure and R&D personnel, Barcelona lies well behind the metropolitan region of Madrid, despite being the dominant region in Catalonia. Between 1990 and 1995 an average of roughly 39 percent of the national R&D expenditure and roughly 36 percent of Spain's R&D personnel were concentrated in Madrid. However, in terms of business R&D indicators, such as industrial R&D expenditure or patents applied for, Barcelona achieves better figures than the metropolitan region of Madrid. A significant contrast is that while Madrid benefits greatly from public-funded R&D, in Catalonia it is the firms themselves that are mainly responsible for R&D activities (see Table 3.5).

Table 3.5 R&D activities in Spain: Input indicators

	R&D Expenditure as % of GDP	R&D Expenditure by Sector (%)			R&D Personnel	by Sector (%)		
		Public	Higher Education	Private	in 1,000	Public	Higher Education	Private
Andalucia	0.54	22	51	27	7.11	25	54	20
Aragon	0.61	25	34	42	1.89	29	40	31
Asturias	0.52	18	48	34	1.27	24	50	24
Baleares	0.12	29	61	8	0.27	30	63	7
Canarias	0.42	30	64	6	1.54	27	68	3
Cantabria	0.49	20	51	28	0.55	16	62	20
Castilla-La M.	0.22	16	22	63	0.65	26	34	38
Castilla y Leon	0.60	8	47	45	3.31	11	56	32
Catalonia (Barcelona)	0.91	10	23	66	13.69	12	30	57
Extremadura	0.31	36	53	11	0.62	40	50	11
Galicia	0.43	23	50	27	2.39	28	50	22
La Rioja	0.25	23	20	57	0.17	35	29	41
Madrid	2.13	31	13	56	25.64	37	22	40
Murcia	0.49	27	48	25	1.18	30	51	19
Navarra	0.87	3	47	50	1.42	6	54	40
Pais Vasco	1.14	2	16	79	5.33	3	24	72
Com. Valencia.	0.48	14	54	31	4.00	16	56	27
Spain	0.87	20	27	52	71.04	24	35	40

3.2 Political and Institutional Framework

Besides the importance of foreign direct investments, the very successful catching-up process achieved by Catalonia is also closely associated with socio-political factors (Bacaria and Borras 1998). Fostered by the Franco dictatorship, there is a strong sense of shared identity in Catalonia. Examples are the long tradition in social movements, the leading role of Catalan nationalist parties in the local and regional governments, the survival and everyday usage of the Catalan language, and the entrepreneurial spirit. The resulting trust and feeling of belonging has helped to build up public policies based on consensus (Bacaria and Borras 1998).

Technology policy under the Franco dictatorship was based on the investment of foreign capital, as mentioned above. At that time foreign firms helped to modernise

Spain's industrial structure with their massive technology imports. So far, however, foreign firms have usually only established production facilities, leaving the R&D activities in their home country or elsewhere. At the same time, the Spanish scientific-technological system has remained isolated. The modernisation of industry has not been accompanied by improvements in the national innovation system. After Franco's death in 1976, Spain faced severe problems in the democratic transition process, which was made even more difficult because of the economic crisis and subsequent restructuring of the economy. As a result, Spain started to develop its own technology policy only at a very late stage.

At the national level it is only since the end of the 1980s that an active technology policy has been in operation with the implementation of a National Plan for R&D. On a three yearly basis this plan, which is developed, co-ordinated and controlled by the CICYT (Comision Interdepartamental de Ciencia y Tecnologia), determines which fields of technology and research are to receive state funds. The projects of state research institutions and universities, as well as joint projects between firms and state research institutions, are funded. In addition, the CICYT grants scholarships to help scientific personnel to improve their qualifications.

The Centre of Industrial Technology Development (CDTI), a national agency related to the Ministry of Industry, plays a leading part in funding research projects within firms. The latter can submit project proposals which are evaluated by the CDTI for financing. After funding has been approved, the firms concerned usually receive interest-free loans. In addition to the CDTI, the Ministry of Industry also funds R&D projects in firms through its own programmes. Within its 'Plan de Actuacion Tecnologica Industrial', the Ministry funds R&D projects in seven fields: electronics and informatics, automation, pharmaceuticals, materials for biotechnology and chemistry, basic industry, infrastructure, and training. At the beginning of the nineties, Catalonia was a beneficiary of the public R&D policies under the national plan, and between 1988 and 1990 it received nearly 22 percent of the overall expenditure. Its participation in industrial programmes was even higher. In 1990, nearly 35 percent of the total project finance volume of the CDTI was directed to Catalonia (Bacaria and Borras 1998) (see Table 3.6).

Due to the centralised political structure, experience with regional technology policy is very limited. Since Franco's death a marked decentralisation process has been taking place, aiming at strengthening regional autonomy. From the very beginning, the Catalan government has been very active in developing its own innovation policies. In addition to R&D funding by the Spanish central government and the European Union (regional funds, mainly STRIDE, and the Community Support Framework for Catalonia), the Catalan government has been trying since the beginning of the eighties to reduce the shortcomings in the economic structure, and to provide support for Catalan firms by means of a large number of funding programmes and measures.

The concrete goal of the Catalan government is to enhance the international competitiveness of firms. The aim of its industrial policy is, on the one hand, to modernise and consolidate existing businesses and, on the other, to fund innovative networks, which are regarded as a necessary precondition for establishing new domestic and foreign businesses in the high-tech field. In order to transform the

Catalan innovation system, which is basically concentrated in the metropolitan region of Barcelona, the Catalan government founded two institutions: CIRIT (Comisio interdepartamental de Recerca i Innovacio Tecnologica), founded in 1980, and CIDEM (Centre d'Informacio i Desenvolupament Empresarial) in 1985. CIRIT is an interdepartmental commission linked to the regional Ministry of Research and Universities, and dedicated to scientific research. CIRIT's most important task is the development of the Catalan research plan, which every four years lays down guidelines for Catalonia's technological and innovation policy.

Table 3.6 Some further characteristics of R&D activities in Spain

	Patents per million	Regional Distribution of Public R&D Funds by: Number of Financed R&D Projects as %			R&D Expenses %	
		Ministry of Industry	CDTI [a]	EU	Ministry o Industry	CDTI [a]
Andalucia	1.9	4.0	3.9	2.0	4.5	3.8
Aragon	6.9	2.4	2.4	1.1	2.9	2.4
Asturias	3.5	1.7	3.0	1.9	1.0	3.6
Baleares	3.0	0.2	0.4	0.2	0.1	0.2
Canarias	1.2	0.3	0.2	0.0	0.2	0.1
Cantabria	4.2	0.7	0.8	0.5	0.3	0.7
Castilla-La Mancha	0.9	1.4	1.0	0.2	1.3	0.9
Castilla y Leon	2.7	2.3	1.6	0.5	1.9	1.7
Catalonia (Barcelona)	20.2	22.2	33.1	16.7	31.9	31.2
Extremadura	0.6	0.1	0.3	0.3	0.1	0.3
Galicia	1.1	2.4	2.6	2.7	1.3	2.9
La Rioja	6.4	1.1	0.6	0.0	0.5	0.4
Madrid	16.4	39.1	29.5	55.7	28.8	32.5
Murcia	2.5	1.1	1.2	0.2	0.7	1.3
Navarra	16.7	2.8	3.4	1.4	3.0	2.9
Pais Vasco	9.3	12.5	9.2	12.4	15.5	10.1
Com. Valenciana	5.9	5.6	6.8	4.4	6.2	5.1
Spain	9.1	100.0	100.0	100.0	100.0	100.0

Note: **a** Centre of Industrial Technology Development (CDTI)

The first research plan, approved in 1993, has the task of complementing the activities of the central government and the EU. In addition to raising the quality of basic and applied research in general, the research plan supports R&D projects within well-defined funding structures (agro-industries, chemistry and health). CIDEM is a regional economic promotion agency, whose objective is to foster industrial and technological development in Catalonia. However, it does not solely deal with Catalan and/or Spanish businesses. In order to encourage foreign businesses to invest in Catalonia, it has opened offices in Germany, Belgium, Japan and the USA.

In addition to the initiatives cited above, mention should also be made of the numerous non university R&D institutions which receive some funding from the

State, and which have a decisive influence on the system of regional innovation. Some of these are research institutions established in Catalonia by the National Public Research Centre (Consejo Superior de Investigaciones Cientificas), others are were founded by the Catalan government. Most of these research institutes are located in the metropolitan region of Barcelona. Among the National Public Research Centre's ten research institutes, which are generally attached to universities, special mention must be made of the National Centre for Microelectronics, Institut de Tecnologia Quimica i Textil, Institut de Ciencia de Materials de Barcelona, Institut de Cibernetica, and the Centre de Investigacion i Desenvolupament. For its part, the Catalan government has founded numerous R&D institutions, mainly laboratories. These provide important services for firms. They include, among others, the General Testing and Research Laboratories LGAI (*Laboratori General d'Assaigs i Investigacions*), which carry out technical tests on products to see that they conform to international norms and standards, the Automobile Research Institute ISIADA (*Institut d'Investigacio Aplicada de l'Automobil*) (applied research in the field of cars, test circuits), the Textile Research and Testing Laboratory of Terrassa, the Supercomputation Centre of Catalonia (Bacaria and Borras 1998).

As far as the funding of R&D activities is concerned, the part played by universities should not be underestimated. Historically, Spanish universities were in general predominantly engaged in teaching. Research was not encouraged and only undertaken by self-motivated individuals and departments. Because of this concentration on teaching and basic research, the relationship between universities and industry has traditionally been weak. New conditions have however fostered scientific research and links with industry. Through the creation of private universities, competition between the institutions has increased significantly. As a result, during the nineties the total number of Catalan universities rose from three to eight, and the traditional dominance of the courses of study in the arts and humanities shifted in favour of courses in engineering and the natural sciences. In Barcelona alone, there are three public universities with a total of just under 150,000 students and roughly 6,500 scientists. The Institutes for Physics and Pharmaceutics at the University of Barcelona, the Institute for Basic Research in Biology at the Independent University of Barcelona, as well as the application-orientated Institutes for Automobile Research and Textile Research at the Polytechnic University in Barcelona are all playing an important role with regard to the stronger orientation towards technology. This shows that the universities are moving towards more applied research and aiming to increase the research contacts with industry. In addition, all the universities have several technology transfer facilities.

One highly prestigious Catalan technology project is the Technology Park situated 30 km from Barcelona in the Valles Occidental. This was opened in 1988 and, in an area covering 120 hectares, provides space for 120 firms. Business services are being provided by means of joint agencies comparable to those found in the newly-established German industrial centres. At the moment, Hispano-Olivetti, which produces computers, the National Microelectronics Centre of the University of Barcelona, already mentioned above, as well as branches of multinational concerns like Sharp and Hewlett Packard, have located in the Valles Occidental, which the Catalan government also calls 'Silicon Valles'. The entire valley has been declared a special industrial zone and generous subsidies are granted to firms setting up there (e.g. investment subsidies of up to 30 percent and exemption from import taxes).

In order to diffuse scientific and technological information in the region, an Institute of Technology was created. In addition, technical conferences and international fairs play an active role supporting learning processes in Catalan firms (Bacaria and Borras 1998). The principal aim of these numerous initiatives, especially those promoted by the Catalan government, is to foster the internationalisation of the Catalan innovation system. The Technology Park and the General Testing and Research Laboratories LGAI (*Laboratori General d'Assaigs i Investigacions*) are good examples of an adaptation to international standards. Besides the activities already mentioned, other initiatives include the Patronat Català Pro Europa, a semi-public agency which provides information on European Union R&D programmes and represents Catalan interests, the cross-border co-operation agreements like Euregio, Four Motors, Arc Mediterrannen and the setting up of the Export Office, COPCA. These all demonstrate a political willingness to overcome the historical weaknesses of the Catalan innovation system. Although there appears to be a strong political and social consensus, there is still much room for improvement in the institutional system. Bacaria and Borras (1998) conclude in their analysis of the Catalan innovation system that the strong competition between employers' organisations hinders the optimal co-ordination of vocational training, applied research, technical services and information, which needs to be more competitive.

Despite these funding programmes and the measures that have already been introduced, shortcomings continue to exist in the innovative ability of Catalan firms. It has not yet been possible to establish innovative networks as intended. A study by Escorsa (1994) on the R&D behaviour of 36 Catalan firms comes to the conclusion that the firms questioned make only very limited use of the services offered by universities, research institutes, technology transfer institutions and other technology agencies. A survey conducted by CIRIT confirms these findings, revealing that most innovative ideas were generated internally within the firms themselves.

Although this conclusion is based on an investigation which involved only a very small number of firms, and therefore does not necessarily indicate a general trend, it underlines the need to undertake a comprehensive quantification of the innovative performance of the Catalan firms. It would also be interesting to discover to what extent an innovation network can be said to exist. Such an investigation would have to take into account all the actors affecting the innovative environment, such as industrial firms and public service companies, as well as public and transfer-relevant R&D agencies (see Molero and Buesa 1996b). It is only in this way that an evaluation can be made of the success of the regional technology policy in the metropolitan region of Barcelona. Among other things, this policy aims to encourage innovation networks. An examination of this kind would provide the information necessary to enable concrete recommendations to be made for action to encourage the establishment of such networks.

3.3 The Manufacturing Sector

A The Survey

The data collection stage of the industrial survey proved to be very difficult. Three weeks after the questionnaires had been sent out, at the beginning of October 1997, the response rate was a disappointing 5 percent. By April 1998, three series of telephone follow-ups to the non-responding firms had raised the rate of returned questionnaires to 15 percent. Although still low, this percentage can be regarded as a success compared with other empirical surveys made in Spain. For example, the EU Community Innovation Survey (CIS) was suspended after two weeks as only 5.8 percent of the firms had replied (Archibugi 1996; Evangelista et al. 1997).

Table 3.7 gives an overview of the representativeness of the sample. It shows that the overall structure of the total population was fairly well covered. The biggest difference is 4.7 percentage points in the textile and clothing industry, which is thus slightly underrepresented. As expected, the response rate of small manufacturing firms was in general relatively low. There is some indication of a bias affecting the key variables of interest, which means that the innovation survey tends to overstate aggregate levels of innovative and networking activities.

A general feature of the sample is that a large proportion of the surveyed firms were relatively small. As in the total population of registered firms, approximately 80 percent had less than 100 employees. With respect to their organisational status, 76 percent were single establishments, usually family firms. The remaining firms operated as branch plants. The number of Catalan firms with main plants in the metropolitan region of Barcelona is rather low. The age structure shows that many firms are quite young: only 30 percent of the firms were established before 1970, whereas 25 percent of the surveyed firms have been in business since the beginning of the nineties.

B Innovation Activities of Manufacturing Firms

The innovative activities of firms can be measured by input, throughput and output indicators. Whereas input indicators quantify the resources devoted to R&D efforts (e.g. R&D personnel, R&D expenses), both throughput and output indicators are related to the outcome of R&D activities. Throughput indicators, generally measured in terms of patents, represent R&D results not yet transferred into marketable products. Output indicators give a picture of the importance of innovation activities for current business. They measure the outcome of innovation activities directly, e.g. through the number of newly introduced or improved products or the share of turnover achieved with such products.

Table 3.7 Response patterns and response rate of responding manufacturers

	Total Number of Registered Firms 1997		Number of Responding Firms 1997		Response Rate [a]
	no.	%	no.	%	
Industry Sector					
Textiles & Clothing	448	16.91	49	12.44	10.94
Food Industry	83	3.13	13	3.30	15.66
Wood, Paper & Printing	431	16.26	49	12.44	11.37
Chemicals, Plastics & Rubber	690	26.04	108	27.41	15.65
Electrical & Optical Equipment	270	10.19	50	12.68	18.52
Basic Metals & Metal Products	349	13.17	60	15.23	17.19
Machinery & Transport	379	14.30	65	16.50	17.15
Employment Size					
≤ 49	1.678	63.32	237	60.77	14.12
50 – 99	487	18.38	81	20.77	16.63
100 – 499	416	15.70	55	14.10	13.22
≥ 500	69	2.60	17	4.36	24.64
Total	2.650	100.00	390	100.00	14.72

Note a number of responding manufacturing firms divided by total number of firms multiplied by 100.

Table 3.8 Selected R&D characteristics of surveyed firms (1994-1996)

	Firms with Continuous On-Site R&D 1997		R&D Personnel Ratio [a]	R&D Expenditure Intensity
Industry Sector		b		
Textiles & Clothing	3	6.0	60.6	4.12
Food Industry	1	2.0	17.0	9.68
Wood, Paper & Printing	4	8.0	15.8	0.75
Chemicals, Plastics & Rubber	14	28.0	76.1	4.48
Electrical & Optical Equipment	8	16.0	107.1	4.69
Basic Metals & Metal Products	7	14.0	54.3	4.21
Machinery & Transport	13	26.0	67.5	3.98
Employment Size				
≤ 49	24	48.0	70.9	5.87
50 – 99	12	24.0	78.1	4.98
100 – 499	13	26.0	33.6	2.24
≥ 500	2	4.0	63.9	4.57

Notes: **a** per 1,000 employees
b percentage of all firms in the corresponding row category

Table 3.8 gives a profile of the surveyed firms utilising three indicators: the presence of continuous on-site R&D facilities, the R&D personnel ratio, and the R&D expenditure intensity. It is evident that firms in the chemical industry, machinery and transport, and electrical & optical equipment rely more strongly on their own R&D efforts than more labour-intensive sectors such as textiles & clothing. The food sector and wood, paper and printing have particularly low values for R&D personnel. Nevertheless, the food sector has a surprisingly high R&D expenditure intensity, demonstrating the willingness to introduce innovations generated outside the firm.

Interestingly, a clear size effect is recognisable. Small firms tend to rely more on in-house R&D facilities than large firms. Nearly half of the small firms with less than 50 employees have continuous on-site R&D facilities. Large firms with more than 500 employees are heavily dependent on R&D transfers from outside, usually from the firm's head office.

While Table 3.8 provides evidence of the input side of the innovation process, Table 3.9 focuses on the output side, measured in terms of the innovation rate (number of new products per 1,000 employees) and the share of turnover accounted for by new or improved products (averaged over 1994-1996). The advantage of output indicators is that they reveal the economic success of firms' innovation efforts.

Table 3.9 Innovation activities of surveyed firms (1994-1996)

	Innovation Rate [a]	Share of Turnover by Product Innovation
Industry Sector		
Textiles & Clothing	53.7	0.65
Food Industry	8.7	0.06
Wood, Paper & Printing	169.4	0.14
Chemicals, Plastics & Rubber	66.2	0.13
Electrical & Optical Equipment	335.8	0.41
Basic Metals & Metal Products	187.0	0.22
Machinery & Transport	6.9	0.46
Employment Size		
≤ 49	232.4	0.25
50 – 99	124.5	0.26
100 – 499	82.6	0.16
≥ 500	3.4	0.37

Note: a denotes number of new products per 1,000 employees.

Both output indicators show that differences between the industrial sectors and firm size are significant. The industrial sectors of electrical and optical equipment, basic metals and metal products, wood, paper and printing are characterised by high numbers of new or improved products. Here the product life cycles seem to be shorter than in sectors like machinery and transport or food industries. However, taking the share of turnover by product innovation the picture is slightly different. This indicator also considers the incremental improvement of existing products, resulting in high shares in the textile and clothing, in electrical and optical equipment and in the machinery and transport sectors. These are sectors which have to respond very quickly to changing market needs.

In addition, the responding firms were classified as innovative or non-innovative according to the share of turnover accounted for by new or improved products, following Malecki and Veldhoen (1993). Firms were therefore considered innovative if product innovations introduced during the past three years comprised more than 20 percent of the yearly turnover. Defined in this way, only 30 percent of the firms could be regarded as innovative (compared with 50 percent in Vienna). Table 3.10 provides a breakdown of these innovative firms by firm size. It shows that the majority of the innovative firms are very small, having less than 50 employees (63 percent of all innovative firms). The textile and clothing, the electrical and optical equipment and the machinery and transport sectors showed a predominance of innovative firms, and together they accounted for 56 percent of all innovative firms.

Table 3.10 Distribution of innovative firms by sector and size (1994-1996)

	Innovative Firms [a]		Distribution by Employment Size (no. employees)							
			Under 50		50 - 99		100 - 499		500 and larger	
	no.	% [b]	no.	% [c]	no.	% [c]	no.	% [c]	no.	% [c]
Industry Sector										
Textiles & Clothing	13	52	10	52	0	0	2	100	1	100
Food Industry	1	14	1	25	0	0	0	0	0	0
Wood, Paper & Printing	10	32	8	38	2	40	0	0	0	0
Chemicals, Plastics & Rubber	10	16	3	36	3	23	2	12	2	33
Electrical & Optical Equipment	13	62	9	69	2	40	2	100	0	0
Basic Metals & Metal Products	9	26	5	20	4	50	0	0	0	0
Machinery & Transport	12	29	7	33	3	21	1	25	1	50
Total	68	31	43	34	14	29	7	23	4	31

Notes: **a** A firm is classified as innovative if product innovations introduced during 1994-1996 comprised over 20 percent of annual turnover;
b Percentage of innovative firms of all firms in the corresponding sector;
c Percentage of innovative firms of all firms in the corresponding sector and size class.

C Networks and Network Formation

As stressed in Chapter 1, R&D activity may be misleading, or at least incomplete, as an indicator of technological capacity, since it does not include network activities, learning, informal R&D and other means of enhancing a firm's knowledge base. Table 3.11 provides some empirical evidence of the role of internal R&D efforts in comparison with external R&D sources, such as external co-operation and the acquisition of licenses. It is very clear from the survey results that the firm's own R&D efforts and the accumulation of production experience are considered the most important sources for product innovation, while only medium importance is attached to external R&D sources. Interestingly, the accumulation of production experience is perceived as being of greater importance than a firm's own R&D efforts (with values of 84 and 81 respectively on the 'scale of importance'). This may indicate that product innovations are not so much a result of institutionalised R&D processes as of product improvements made during the production process. Despite this, outside R&D sources are still relevant.

Table 3.11 Ways of generating product innovations by index of importance (1994-1996)

	Generation of Product Innovations[a] by:			
	Own R&D	**Accumulation of Production Experience**	**External Co-operation**	**Acquisition of Licenses**
Industry Sector				
Textiles & Clothing	81	89	53	22
Food Industry	88	90	40	20
Wood, Paper & Printing	71	81	59	41
Chemicals, Plastics & Rubber	87	79	59	39
Electrical & Optical Equipment	88	85	42	26
Basic Metals & Metal Products	76	92	53	24
Machinery & Transport	77	82	55	25
High Technology [b]				
Employment Size				
≤ 49	79	85	52	29
50-99	82	86	54	30
100-499	87	80	58	34
≥ 500	82	76	68	38
Total	81	84	54	31

Notes: **a** Index of importance measured on a scale from 0 to 100 as follows: 100 = all firms give maximum importance, 0 = all firms give minimum importance;
b High technology sectors defined in the sense of Hatzichronoglou (1997).

External co-operation, in particular, is considered fairly important, while the acquisition of licenses appears to be used to a much lesser extent as a strategy for absorbing external knowledge (with values of 54 and 31 respectively). It is important to note that the sector variations are not very pronounced. The food industry, and electrical and optical equipment seem to attach rather more importance to own R&D and tend to rely less on external co-operation networks compared to the other sectors.

The breakdown of manufacturing firms by size classes shows that large firms with more than 500 employees evaluate the importance of external co-operation much more highly than smaller firms (an importance index of 68 compared to 52 for small firms with less than 50 employees). They also devote more resources to the acquisition of licenses (importance index of 38 and 29 respectively).

The importance of external co-operation is revealed more clearly in Table 3.12. Manufacturing firms were asked with whom and in which phase of the innovation process they co-operate, and where the co-operation partner is located. Innovation here was seen in a broader sense, including product, process and organisational innovations. The following five types of co-operation partner were distinguished: customers, manufacturing suppliers, other manufacturers (e.g. competitors), producer service suppliers and research institutions (see Chapter 1). Overall, 79 percent of the sample firms were involved in external network activities with at least one co-operation partner.

Table 3.12 Sectoral and size characteristics of external innovation co-operation (1994-1996)

	Networking Firms	Percentage of Networking Firms of all Firms in the given Category	Average Number of Connections
Industry Sector			
Textiles & Clothing	19	76.0	22
Food Industry	4	57.1	21
Wood, Paper & Printing	24	77.5	17
Chemicals, Plastics & Rubber	53	85.5	35
Electrical & Optical Equipment	21	87.5	24
Basic Metals & Metal Products	26	74.3	25
Machinery & Transport	32	78.0	33
Employment Size			
≤ 49	97	75.8	22
50-99	42	85.7	28
100-499	28	87.5	37
≥ 500	10	76.9	53
Total	179	79.2	27.8

This high degree of external networking would seem to contrast with the evidence shown in Table 3.11, but it is important to note that this observation is based on *actual* co-operation links, while Table 3.11 refers to the firms' *perception* of the importance of different R&D strategies for their product innovation activities. The results show that the firms surveyed tend to underestimate the role of external co-operation, which in fact seems to be much more important than perceived.

Variations can be observed between sectors. The highest proportions of networking firms were recorded in electrical and optical equipment, and the chemicals, plastics and rubber sector (88 percent and 86 percent respectively). The machinery and transport sector, and wood, paper and printing were slightly below the average, while the lowest percentage was found in the food sector. Co-operation intensity measured by the average number of connections was highest in chemicals, plastics and rubber, and in machinery and transport. With respect to firm size, medium-sized firms (from 50 to 499 employees) were shown to be much more involved in network activities than small and large firms. Surprisingly, the co-operation intensity of large firms is significantly higher than that of the smaller size classes.

Table 3.13 provides empirical evidence of the five types of networks identified in Chapter 1: customer networks, manufacturing supplier networks, producer service supplier networks, producer networks, and co-operation with research institutions. 52 percent of the manufacturing firms co-operate with producer service providers, 50 percent with customers and 43 percent with manufacturing suppliers. Horizontal co-operation activities appear to be relatively unimportant. The low level of co-operation with research institutions is especially significant to note, in view of the enormous efforts made by national, regional, and EU programmes in promoting university-industry collaboration.

Co-operation is more common in the early stages of the innovation process. During the early stages (pre-competitive stage), which are characterised by the exchange of information, the identification of new ideas and joint R&D efforts, the main intention of manufacturing firms is to explore new technological possibilities. Only with customers, and to a lesser extent with manufacturing suppliers, is co-operation more balanced through all the innovation phases. Besides intensive co-operation in the early stages, the responding firms collaborate intensively in prototype development, pilot application and market introduction with their customers and manufacturing suppliers.

A further aspect which emerges is that there are differences between size classes of innovative firms in regard to their co-operation partners. Whereas large firms are relatively active in co-operating with research institutes, smaller firms tend to co-operate with manufacturing customers and, to a lesser extent, with producer service firms. Medium sized innovative firms clearly dominate the co-operation with producer service firms.

Table 3.13 Network activities of manufacturing firms (1994-1996)

		Customer Networks	Manu-facturing Supplier Networks	Producer Service Supplier Networks	Producer Networks	Co-operation with Research Institutions
Pre-Competitive Stage						
Information Exchange	**a**	104	86	97	31	33
	b	46.0	38.1	42.9	13.7	14.6
Identification of New Ideas	**a**	96	76	83	30	26
	b	42.5	33.6	36.7	13.3	11.5
Research and Development	**a**	74	68	81	25	31
	b	32.7	32.7	30.1	11.1	13.7
Competitive Stage						
Prototype Development	**a**	81	72	64	19	30
	b	35.8	31.9	28.3	8.4	13.3
Pilot Projects	**a**	74	58	57	15	21
	b	32.7	25.7	25.2	6.6	9.3
Market Introduction	**a**	79	44	39	18	13
	b	35.0	19.5	17.3	8.0	5.8
Total	**a**	114	96	118	40	46
	b	50.4	42.5	52.2	17.7	20.4

Notes: **a** denotes the number of such network activities of manufacturing firms;
b denotes the percentage of networking manufacturing firms.

As discussed above, in-house R&D efforts seem to be more important in the metropolitan region of Barcelona than external collaboration as a source of product innovations (see Table 3.8), but differences can be observed between innovative and non-innovative firms. Table 3.14 demonstrates that the former tend to be more geared to external co-operation activities than non-innovative firms. Horizontal linkages with research institutions or other manufacturing firms are maintained by both innovative and non-innovative firms, in a similar manner, but both are on a very low level.

While co-operation with research institutions is seen, above all, as an opportunity to enter new technological fields and to acquire know-how, in the case of co-operation with other manufacturing firms, the most important motive seems to be risk reduction. As far as co-operation with producer service providers is concerned, it is not possible to detect any clear picture. Risk reduction, entering new technological fields, financial resources and the acquisition of funds seem to be reasons of roughly equal importance for entering into co-operation with producer services, while the acquisition of know-how is relatively unimportant.

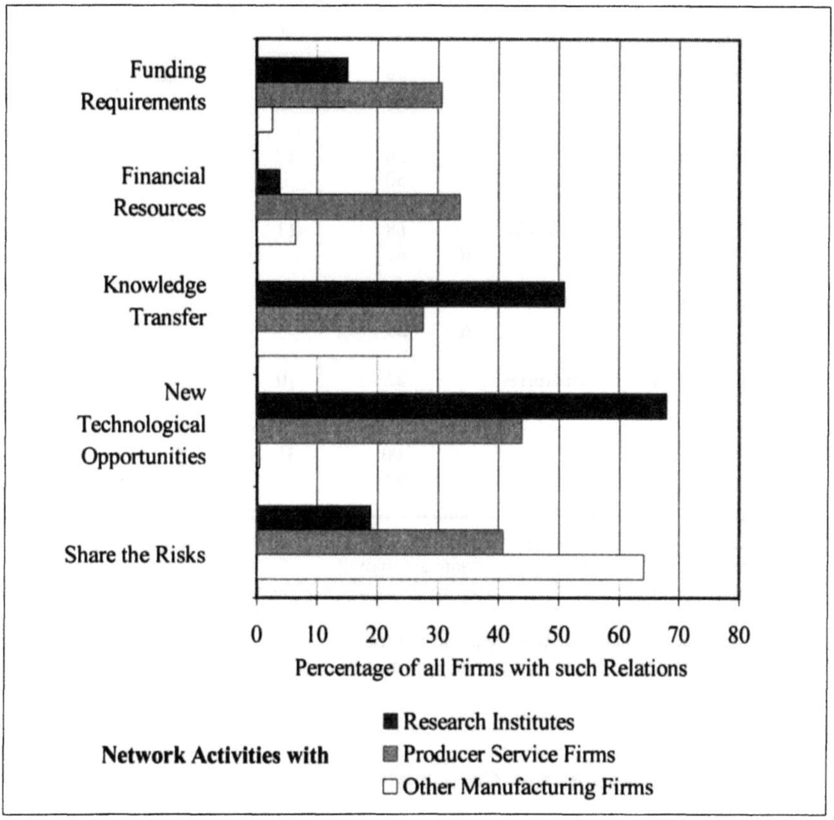

Fig. 3.2 Motives for networking activities (1996)

The problems raised by innovation co-operation appear to depend on the partner concerned (see Fig. 3.3). While in co-operative relations with research institutes, co-ordination difficulties were cited as the most significant problem (46 percent of manufacturing firms mentioned such problems), in the case of co-operation with producer service providers, it was the problem of overruns in budgeted cost (50 percent). The lack of schedule effectiveness was seen as the greatest problem in co-operation with other manufacturing firms. Interestingly, rivalry factors like unintended knowledge drain, confidential relation/secrecy or loss of independence were not considered serious hindrances to co-operation with other firms. Differences in capability also appear not to be a major problem in collaboration activities.

Table 3.14 Do innovative firms network more?

		Innovative Firms [a]		Non-Innovative Firms	
Network Type		b	c	b	c
Customer Networks		68	16	59	15
	d	61		47	
Manufacturing Supplier Networks		56	12	54	15
	d	50		43	
Producer Services Supplier Networks		68	11	65	10
	d	61		53	
Producer Networks		25	11	23	11
	d	23		19	
Co-operation with Research Institutes		25	10	28	8
	d	23		23	
Networks - General		100	31	100	28
	d	90		81	

Notes: **a** firms are classified as innovative if product innovations introduced during 1994-1996 comprised more than 20 percent of annual turnover;
b denotes the percentage of manufacturing firms with such networking activities of all manufacturing firms;
c denotes the number of network relations per firm;
d denotes the percentage of manufacturing firms with such networks of all firms in the given category.

D The Location of Innovation Co-operation Partners

Distance seems to be a factor which has an important influence on the distribution of co-operation partners. The metropolitan region of Barcelona hosts by far the largest number of co-operation partners. 74 percent of all networking firms maintain innovation-led co-operative relationships with firms and research institutes within this region (see Table 3.15). It would seem to confirm that face-to-face contacts can facilitate interpersonal relationships and favour the mutual trust which serves as a basis for business co-operation. 55 percent of all networking firms co-operate with partners in the rest of Spain, while only 47 percent have partners located in the rest of Catalonia.

The geographical distribution of co-operation partners is clearly dominated by local and national links than international ones. The European Union is the main non domestic source of innovation, while other macro regions e.g. USA, Latin America or the rest of the world are much less significant. This means that local and to a lesser extent national sources for innovation are crucial for manufacturing firms in the metropolitan region of Barcelona, indicating some weaknesses with respect to the internationalisation of R&D activities.

The spatial distribution of co-operation partners for innovative and non-innovative firms is shown in Table 3.16. Overall, the reliance on international knowledge sources is very limited. Innovative firms report networking activity which is three to four times more intense in all geographical areas than non-innovative firms. It seems that innovative manufacturing firms are much better connected and integrated into local, regional and national innovation networks than non-innovative firms. The co-operation partners of non-innovative firms are concentrated at the local, regional and national levels.

Table 3.15 Geographical characteristics of external innovation co-operation (1994-1996)

	Co-operation Partners	Percentage of Networking Firms With Co-operation Partners Located in the Given Category	Average Number of Connections
Location of Co-operation Partners			
Metropolitan Region	168	74.3	8.0
Rest of Catalonia	107	47.3	8.0
Rest of Spain	124	54.9	8.0
European Union	96	42.5	8.0
USA	42	18.6	8.0
Latin America	33	14.6	9.0
Rest of the World	38	16.8	9.0
Total	179	79.2	27.8

Table 3.16 Metropolitan, national and international connections (1994-1996)

	Share of Firms with Network Relations of All Firms	
	Innovative Firms [a]	Non-Innovative Firms
Metropolitan Region	86	28
Rest of Catalonia	59	19
Spain	63	20
European Union	46	16
USA	16	7
Latin America	16	6
Rest of the World	16	6

Note: a firms are classified as innovative if product innovations introduced during 1994 -1996 comprised more than 20 percent of annual turnover.

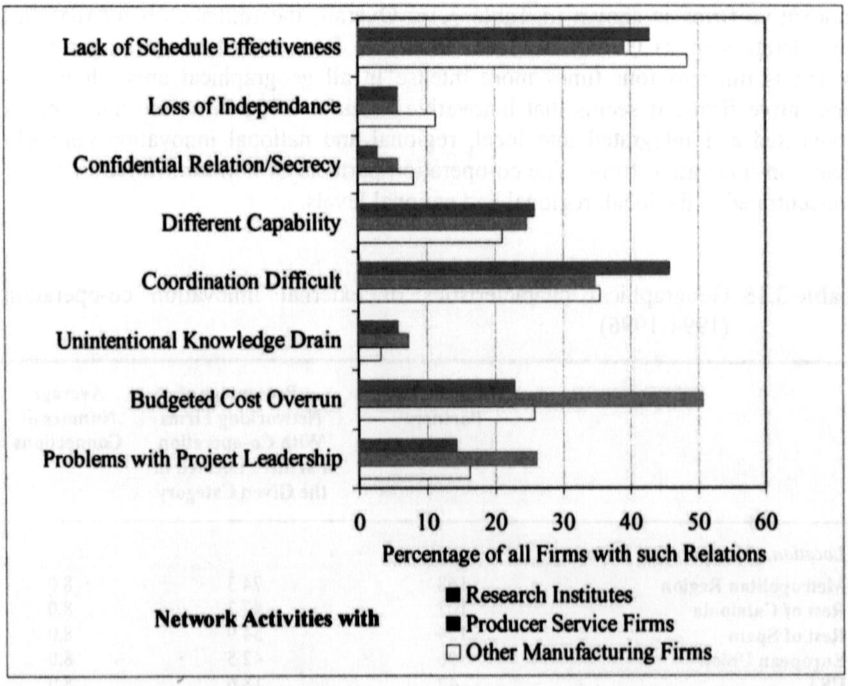

Fig. 3.3 Problems with network activities (1994-1996) (as a percentage of all networking firms)

3.4 The Producer Services Sector

A Some General Features

In knowledge-based economies, producer service firms are becoming crucial in industrial R&D. In advanced economies the service sector is by far the most dynamic sector, having the highest employment increases and being highly innovative. The importance of its role for innovation processes in manufacturing firms is being increasingly acknowledged. The increasing complexity of innovation processes and the tendency of manufacturing firms to concentrate on their core competencies, are leading to a reduction of in-house R&D activities. In order to obtain the necessary technological input to remain competitive, manufacturing firms outsource R&D to specialised producer service firms. These firms have a very intensive interaction with their clients, since they have to provide highly customised products. Their role is to bring together their expertise with the precise requirements of the client.

The producer services sector plays a crucial role within any system of innovation. Following Strambach (1999), producer service firms essentially fulfil four functions:

- Transfer of knowledge: producer service firms provide services in the form of technological and/or management know-how;
- Knowledge integration: as a result of the increased complexity of the customer needs, producer service firms have to integrate and combine different fields of knowledge;
- Adaptation of existing knowledge: producer service firms are able to adapt existing knowledge to specific customer needs;
- Production of new knowledge: through the intensive co-operation with customers and through own innovative capabilities, producer service firms collect and rearrange knowledge into new products.

Compared to the Austrian and the Swedish systems of innovation, in Spain the producer services sector plays a far more modest role (OECD 1999). The purpose of this section is to provide some empirical evidence on the sector, and to characterise the nature and extent of its innovation support activities for the manufacturing sector in the metropolitan region of Barcelona.

Table 3.17 indicates response patterns by sector and size. Compared with the Vienna survey, the distribution of sample firms by sector reflects the pattern for the overall population much less closely. The computer software sector is considerably over-represented, while technical and business consultancy firms are both under-represented.

B Innovation in Producer Service Firms

It is not only manufacturing firms which have to introduce innovations in order to achieve and maintain competitiveness, producer service firms too are facing an increasingly competitive environment. In order to support innovation activities, they have to show a certain degree of innovativeness themselves. Innovations in producer services are classified under service innovations, and are comparable to product innovations in manufacturing, when they involve the substantial improvement of existing services or the development of new ones (e.g. new software development, improved testing methods), or organisational innovations which include new or substantially improved methods of production and/or of fulfilment of customers' needs.

Table 3.18 provides empirical evidence of the innovativeness of producer service firms by sector and size class. Two-thirds of the sample firms introduced new services during the period of observation. Organisational innovation occurred in 59 percent of the sample firms. These figures indicate a certain concern for the improvement or new development of services offered. Significant variations can be observed between sectors. Computer software and business consultancy tend to be more innovative than market research and advertising, or technical consultancy. This holds true both for new services and organisational innovation.

Table 3.17 Response patterns and response rate of responding producer service firms

	Total Number Registered Firms 1997		Number of Responding Firms 1997		Response Rate [a]
	no.	**%**	**no.**	**%**	
Producer Service Provider Categories					
Computer Software	174	29.10	40	38.10	22.99
Technical Consultancy	126	21.07	16	15.24	12.70
Business Consultancy	191	31.94	29	27.62	15.18
Market Research and Advertising	107	17.89	20	19.04	18.69
Employment Size					
≤19	319	53.26	38	37.25	11.91
20-49	185	30.88	43	42.16	23.24
50-249	80	13.36	15	14.71	18.75
≥250	15	2.50	6	5.88	40.00
Total	599	100.00	102	100.00	17.03

Note: a number of responding producer service firms divided by total number of registered firms multiplied by 100.

Table 3.18 Innovation activities of producer service firms (1994-1996)

	New Services		Organisational Innovation	
Producer Service Provider Categories	**a**	**b**	**a**	**b**
Computer Software	28	70	25	63
Technical Consultancy	9	56	9	56
Business Consultancy	19	66	16	55
Market Research and Advertising	15	75	11	55
Employment Size of Producer Service Providers				
≤19	22	58	20	53
20-49	32	74	28	65
50-249	11	73	9	60
≥250	4	67	3	50
N	69	68	60	59

Notes: **a** denotes number of producer service providers that introduced new products during 1994-1996;
b denotes the percentage of such producer service providers of all producer service providers in the corresponding row category.

Table 3.19 characterises the innovation activities of producer firms in terms of two input measures: R&D expenditure (R&D expenditure as a percentage of the annual turnover) and the R&D personnel ratio (R&D personnel as a percentage of total number of employees). The table shows that there are considerable differences between the sectors and also size classes. In relation to expenditure, computer software firms are the most R&D intensive, but if we consider the personnel ratio, technical consultancy firms appear to be the most innovative. With respect to firm size, R&D expenditure is most intense in the smallest businesses, those with less than 20 employees, whereas the next size class, 20-49, has the highest ratio of R&D personnel.

When we apply the output indicator, i.e. the percentage of turnover accounted for by new services, the above results are confirmed. The high R&D expenditure in computer software firms seems to be channelled into the production of new services, since 26 percent of turnover is generated by newly introduced services. However, the turnover generated by innovative activities is much lower in the other producer service sectors. With respect to firm size, businesses with 20 to 49 employees tend to be the most innovative (see Table 3.20). But again, care should be taken in interpreting these figures. These relatively low R&D indicator values may be a result of the difficulties faced in responding to the questionnaire. Some service firms complained that it is not easy to distinguish between a customised product and a new service.

Table 3.19 R&D activities of producer service firms (1994-1996)

	R&D Expenditure Intensity [a]	R&D Personnel Ratio [b]
Producer Service Provider Categories		
Computer Software	8.3	15.4
Technical Consultancy	4.9	17.7
Business Consultancy	5.5	12.5
Market Research and Advertising	3.1	6.5
Employment Size		
≤19	8.7	27.2
20-49	5.5	28.7
50-249	3.7	15.6
≥250	2.7	3.7
Total	7.2	10.8

Notes: **a R&D expenditure as percentage of annual turnover;**
b R&D personnel as percentage of number of employees.

Table 3.20 Turnover by product innovations (1994 – 1996)

	Percent of Turnover by New Services
Producer Service Provider Categories	
Computer Software	26
Technical Consultancy	10
Business Consultancy	15
Market Research and Advertising	10
Employment Size	
≤ 19	17
20-49	20
50-249	17
≥250	2
Total	17

As for the manufacturing sector, we considered producer service firms to be innovative if, during the last three business years, over 20 percent of turnover was accounted for by new or improved services. According to this definition, the proportion of innovative firms is almost 30 percent, almost as high as for manufacturing firms. Computer software, with 57 percent, is the most innovative sector, while technical consultancy, and market research and advertising, with 10 percent each, are the least. The size breakdown shows that smaller producer service firms tend to be slightly more innovative than larger ones.

The survey confirmed that the main reason carrying out for innovative activities is to increase competitiveness. Table 3.21 shows clearly that for most firms the main objective of innovation is to improve their market potential. This can be achieved by improving the quality of services offered, increasing flexibility or extending applications for existing services. Interestingly, an improvement in goodwill, for example through image improvement, is seen as being much more important than cost reduction as an objective. This overall pattern is valid for all the sectors.

Table 3.21 Objectives of innovation activities (1994-1996)

	Cost Reduction		**Improved Goodwill Values**		**Improved Market Potential**	
Producer Service Provider Categories	**a**	**b**	**a**	**b**	**a**	**b**
Computer Software	12	29	30	71	42	100
Technical Consultancy	8	36	14	64	19	86
Business Consultancy	10	30	25	76	29	88
Market Research and Advertising	6	30	18	90	15	75
Total	36	32	87	78	105	94

Notes: Only those firms considering the effects as important are counted.
a denotes the number of producer service firms;
b denotes the percentage of producer service firms with innovations in the corresponding sector.

As mentioned above, producer service firms have to provide customised products which are highly complex. But are these firms able to depend on their own knowledge base or do they have to use external knowledge sources? Table 3.22 demonstrates that hardly any of the producer firms surveyed rely exclusively on their own R&D capabilities. The preferred knowledge sources are suppliers, other producer service firms and customers, in the form of manufacturing clients. Research institutes play only a minor role as external knowledge sources.

C Innovation Support for Manufacturing Firms

Manufacturing firms were the most important clients for the producer service firms surveyed. 77 percent of the average turnover was contributed by manufacturers, followed a long way behind by other service firms with 15 percent. Table 3.23 shows that this overall pattern is valid for the computer software, technical consultancy and market research and advertising sectors. In contrast, business consultancy firms generate their highest proportion of turnover from other service firms. The public sector is an important client only for technical consultancy firms, and market research and advertising firms.

Table 3.22 External knowledge sources of producer service firms for own innovation (1994 - 1996)

	Number of Producer Service Firms Using External Knowledge Sources	Percentage of Producer Service Firms of all Firms in the Row Category
Co-operation Partners		
Manufacturing Clients	40	34
Direct Competitors	33	28
Other Service Firms	63	53
Other Businesses	31	27
Universities/Research Institutes	26	22
Producer Service Provider Categories		
Computer Software	38	95
Technical Consultancy	16	100
Business Consultancy	29	100
Market Research and Advertising	18	90
Employment Size		
≤ 19	41	100
20-49	41	100
50-249	12	80
≥250	6	100
Total	101	86

Table 3.24 underlines the contribution made by producer service firms in supporting the innovation activities of their manufacturing clients. The majority (77 percent) of the producer service firms surveyed confirmed that they support manufacturing firms in their innovation efforts. Within this overall picture, differences between the sectors and size classes can be observed. It is shown that the technical consultancies, as well as the market research and advertising sector, are very strongly oriented towards manufacturing firms, with respective shares of 94 percent and 80 percent. With respect to firm size, high co-operation intensity with manufacturing firms was reported by the producer service firms with more than 50 employees. 87 percent of the firms with 50 to 249 employees, and 100 percent of the firms with more than 250 employees indicated that they gave innovation support to manufacturing firms.

Table 3.23 Distribution of turnover of producer service firms by customer groups (1994 – 1996)

	Producer Service Provider Categories									
	Computer Software		**Technical Consultancy**		**Business Consultancy**		**Market Research and Advertising**		**Total**	
	a	**b**	**a**	**b**	**a**	**b**	**a**	**b**	**a**	**b**
Average Turnover	819.6		54.2		106.3		157.0		1,139.2	
Customer Groups										
Manufacturing	717.3	87	31.7	59	37.3	35	85.4	54	871.8	77
Service Firms	76.7	9	9.2	17	50.6	48	37.8	24	173.8	15
Public Sector	22.5	3	12.8	23	6.6	6	29.7	19	71.6	6
Private Persons	5.6	1	0.5	1	11.8	11	4.1	3	22.0	2

Notes: **a** denotes turnover in million Euro;
b denotes turnover in percentage of average firm's turnover (average over category).

The questionnaire for producer service firms distinguished four different types of innovation support for manufacturing firms: providing new services, organisational innovations (in the sense of process innovations), improvements in business organisation and the opening of new sales markets. The most important type of innovation support, provided by 57 percent of all responding firms, was organisational innovation, followed by improvement of business organisation (52 percent of firms). There are different specialisation patterns in different sectors. While computer software and business consultancy primarily offer help in process and organisational innovations, technical consultancies mainly support process innovations. The sector of market research and advertising is not only engaged in opening up new sales markets, but they also actively influence the product innovation decisions of manufacturing firms. This shows the importance of this sector in evaluating the future economic outcome of new products.

With respect to co-operation with manufacturing firms, producer service firms, which tend to be small in size, do not rely only on their own capabilities. A single producer service firm is seldom able to fulfil the increasingly complex needs of clients. The result is that such firms are tending to work jointly with external institutions. Through external co-operation, the firm gains access to external knowledge inputs and can enlarge its knowledge base. Table 3.25 shows that to serve the manufacturing clients, co-operation with other producer services is crucial. 63 percent of the surveyed firms involved in manufacturing networks have connections with other producer service firms. Research institutes, however, play a minor role in providing knowledge to producer service firms (23 percent).

Table 3.24 Innovation support for manufacturing clients by producer service firms (1994-1996)

	Producer Service Firms Supporting Innovation	
	a	**b**
Producer Service Provider Categories		
Computer Software	30	75
Technical Consultancy	15	94
Business Consultancy	20	69
Market Research and Advertising	16	80
Employment Size of Producer Service Providers		
≤ 19	26	68
20-49	34	79
50-249	13	87
≥ 250	6	100

Notes: **a** denotes the number of producer service firms that supported manufacturing in their product/process innovation activities (1994 -1996);
b denotes the percentage of innovation supporting firms of all responding firms in the corresponding sector and size class.

Table 3.25 External co-operation of producer services providers with other services and research units to support manufacturing innovation (1994-1996)

	Co-operation with other Producer Service Firms		Co-operation with Research Units	
	a	**b**	**a**	**b**
Producer Service Provider Categories				
Computer Software	16	70	5	22
Technical Consultancy	9	69	6	46
Business Consultancy	7	50	2	14
Market Research and Advertising	8	57	2	14
Employment Size of Producer Service Providers				
≤19	13	62	7	33
20-49	19	73	4	15
50-249	7	58	1	8
≥ 250	1	25	2	50
Total	40	63	15	23

Notes: **a** denotes the number of co-operation partners;
b denotes the percentage of co-operating producer service providers of all producer service firms supporting manufacturing in the corresponding sector and size class.

Table 3.26 Innovation support for manufacturing clients by producer service firms in different stages of the innovation process (1994-1996)

	Pre-Competitive Stage						Competitive Stage					
	Information Exchange		Identification of New Ideas		Research and Development		Prototype Development		Pilot Projects		Market Introduction	
	a	b	a	b	a	b	a	b	a	b	a	b
Producer Service Provider Categories												
Computer Software	21	88	21	88	24	100	21	88	23	96	9	38
Technical Consultancy	14	78	16	89	15	83	11	61	12	67	7	39
Business Consultancy	16	100	16	100	16	100	4	25	11	69	9	56
Market Research and Advertising	10	71	11	79	10	71	7	50	6	43	12	86
Employment Size of Producer Service Providers												
≤ 19	20	80	21	84	19	76	14	56	19	76	13	52
20-49	29	97	29	97	30	100	19	63	19	63	12	40
50-249	8	67	8	67	10	83	8	67	9	75	9	75
≥ 250	2	50	4	100	4	100	2	50	4	100	2	50
Total	59	83	62	87	65	92	43	61	51	72	36	51

Notes: **a** denotes the number of firms indicating innovation support in industry;
b denotes the percentage of innovation supporting firms of all responding firms in the corresponding row category.

The survey has so far provided some empirical evidence on the importance of producer services in supporting innovation activities in manufacturing firms. Table 3.26 shows in which stage of the innovation process this support is given. Co-operation with manufacturing firms tends to take place in the early stages of the innovation process. However, large variations can be observed between sectors. Computer software firms are involved in supporting manufacturing firms in nearly all stages of the innovation process, except market introduction. Technical consultants follow this same pattern. Business consultancy tends to be very strong in the pre-competitive stage of the innovation process. Market research and advertising firms are also engaged in market introduction.

Large sectoral differences are evident. The more technically-oriented sectors are far more actively involved in networks with other producer services than the more business-orientated sectors. Nearly 70 percent of computer software and technical consultancy firms are connected with other producer services. Co-operation with research institutes is undertaken mainly by technical consultancy firms.

D Location of Manufacturing Clients, Co-operation Partners and Innovation-Relevant Sources

The analysis of the location of co-operation partners has given some indications of the importance of spatial proximity in Barcelona's metropolitan innovation system. Producer service firms were asked to rank the most important factors for successful co-operation with manufacturing firms. Frequent personal contacts (33 percent) and good knowledge of a client's industry (29 percent) were cited. Although spatial proximity can be very helpful for these two factors, only 14 percent of responding firms considered it very important for successful co-operation. However, a look at the location of manufacturing clients gives a different picture. The vast majority of manufacturing clients are located in the metropolitan region of Barcelona or in the rest of Catalonia.

Fig. 3.4 shows clearly that spatial proximity has a more important role than perceived by the firms surveyed. Some sectoral differences can be distinguished. Whereas the more business-orientated sectors – business consultancy, and the market research and advertising – concentrate predominantly on manufacturing clients, the more technically orientated sectors – computer software and technical consultancy – have manufacturing industrial clients at the European and global scales. But even for the latter sectors, the largest share of manufacturing clients is to be found in the domestic market. However, in an international perspective, technical consultancy seems to be the most competetive service sector.

Not only is the geographical distribution of manufacturing clients concentrated chiefly at the regional and national level, but also the co-operation partners in innovation projects supporting manufacturing firms. Whereas the majority of the producer service partners are located in the metropolitan region of Barcelona or in the rest of Spain, partners in research institutes are exclusively located in the metropolitan region of Barcelona.

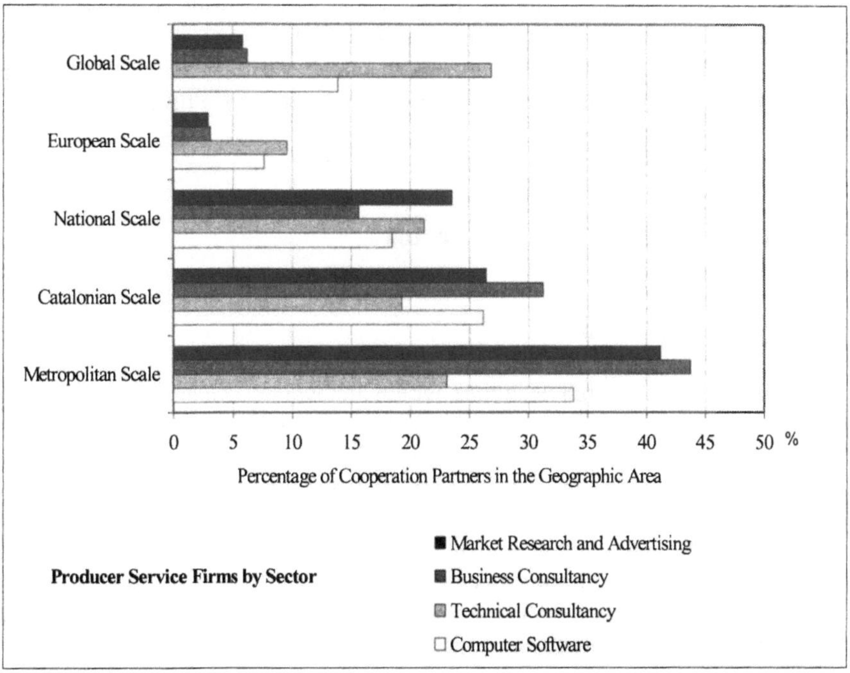

Fig. 3.4 Location of manufacturing clients supported in innovation activities by producer service firms (1994-1996) (percentage of clients in each geographical area for each sector)

Table 3.27 Location of producer service providers' co-operation partners (1994-1996)

	Co-operation of Producer Service Units with			
	Other Service Firms		Research Units	
	a	b	a	b
Metropolitan Region	37	60	15	24
Catalonia	17	37	3	7
Rest of Spain	22	61	3	8
European Union	10	83	3	25
Rest of the World	14	52	7	25
Total	40	63	15	23

Notes: **a** denotes the number of co-operation partners;
b denotes the percentage of co-operating producer service firms out of all producer service firms supporting manufacturing in the corresponding sector and size class.

The above provides an indication of the importance of face-to-face contacts, which is related to the tacit nature of transferred knowledge. It is interesting that apart from local research institutes, the producer service firms surveyed tend to co-operate far more with international institutes, both inside and outside the EU, than those in the rest of Spain. But overall, they make far less use of research units for enlarging their knowledge base than other producer service firms.

3.5 The Science & Research Sector

A Some General Features

The main role of research institutes within innovation systems is in offering scientific knowledge to interested firms, training highly skilled workers and, through their gateway function, ensuring access to international knowledge sources. They are also a driving force for structural change. Via university spin-offs, they can help a regional economy to develop a high-tech sector with high growth and employment potential. Until now, the role of research institutions has generally been examined through a case study approach. There has been a lack of comprehensive analysis of the importance of research institutions for innovations in manufacturing firms, producer services or other research institutes.

Not only are the framework conditions for businesses becoming more competitive. Public research institutes, and especially universities, also face new challenges. This is due to:

- the move from an elite to a mass system of higher education during the last three decades,
- a new emphasis on lifelong learning,
- the pursuit of greater efficiency in public funding,
- the rise of new modes of knowledge production and distribution outside universities (in-house universities of large firms, private universities),
- increasing international competition through long-distance programmes offered by first-class universities like Stanford, Oxford, Cambridge, Harvard, MIT, etc.) via the Internet (Charles and Goddard 1997).

Universities and research institutions have to respond to these changes in order to be able to attract students and funding. At the same time, local economic actors are showing increased interest in their regional universities and other research institutes, as they recognise that these could play an important role by transmitting innovative impulses to the local economy. They acknowledge that research institutes can be the starting point and/or source of diffusion for locally relevant knowledge, especially for SMEs. Mansfield and Lee (1996) showed that firms in regions with an excellent research infrastructure have a statistically higher prospensity to be early adopters of the results of research efforts conducted by the local research institutes.

Besides the provision of new knowledge for businesses through academic research activities, research institutions also provide knowledge in the form of a skilled workforce through university-leavers and through their own employees. University spin-offs may help to initiate or foster a structural change, reducing regional innovation deficits. Moreover, research institutions offer access to international knowledge and information sources. As gateways to global information resources they can provide local businesses with external knowledge (Fritsch and Schwirten 1998; Charles and Goddard 1997).

The importance of research institutions for regional development is emphasised by Kanter (1995). Her argumentation is very much in line with the conceptual framework introduced in Chapter 1. In the future, only businesses that can meet international standards and that are linked into global networks will be successful. For cities or regions, Kanter predicts that only those which succeed in linking local firms to global networks will prosper. Businesses can meet these challenges if they continually elaborate new concepts, increase their competence and are connected to other actors. With respect to regions, Kanter concludes that world class places can help grow these assets by offering innovative capabilities, production capabilities, quality skill, learning, networking and collaboration (Kanter 1995). In this perspective, research institutions are a crucial locational asset within the globalised economy (Kanter 1995; Charles and Goddard 1997).

However, the potential regional economic impact of research institutions is closely connected to the local conditions for absorbing the innovative capabilities. Without favourable local conditions the regional impact is limited. Koschatzky (1995) stressed that the regional impacts of research institutes depend on whether these institutions provide complementary knowledge inputs to the internal innovation efforts of businesses. Positive effects are most likely if the innovation activities of businesses occur in similar technology fields (Feldman 1994; Becker and Peters 1997).

The data on research institutions in the metropolitan region of Barcelona was collected from a total of 424 research units within universities and other public research institutes. In order to obtain a detailed insight into the research and network activities of the research institutions, a questionnaire was sent to the departmental level or, in the case of very large departments, to the individual research unit. The total number of research institutions involved was around 54. Research groups from the Universidad de Barcelona, Universidad Autonoma de Barcelona and Universidad Politecnica de Catalunya, from the newly created Universidad Pompeu Fabra and the private Universities La Salle, Esade, and Ramon Llull were included in the sample. Other bodies participating in the survey were public research institutes from the Central, Catalan and Local Government levels, such as the Centre d'Investigacio i Desenvolupament, the Centre de Supercomputacio de Catalunya, the Centre Nacional de Micro-electronica belonging to the national research system, the Centre Catala de Plastic, the Laboratori General d' Assaigs i Investigacions, the Centre de Supercomputacio de Catalunya, the Institut Catala de Tecnologia linked to the Catalan government, and the Fundació Barcelona Centre de Disseny and the Institut de Bioquimica Clinica at the local level.

Table 3.28 presents the responses broken down into the major fields of science & research: architecture; construction and surveying; biology, chemistry and medicine; mathematics, informatics and physics; electrotechnology and mechanical engineering; economics, social and geosciences. 148 research institutes replied to the questionnaire, making the response rate 35 percent higher than in the manufacturing and producer services samples. Overall, the sample is representative of the total population in which there is a clear specialisation pattern. The predominant research fields are mathematics, informatics and physics, followed by electrotechnology and mechanical engineering.

Table 3.29 provides an overview of the main activities of the research institutes surveyed. 54 percent of the total time budget of all scientific staff was devoted to research, mostly applied research. Teaching, with 29 percent, occupies a large part of the total time budget. This relatively high proportion is a result of the historical orientation of research institutes towards teaching. The rest of the time is spent on transfer and other tasks (i.e. representative duties, administration). Between the science fields there are significant differences with respect to basic research. The proportion of basic research is highest in the mathematics, informatics and physics. The field of electrotechnology and mechanical engineering devotes the largest proportion of its human resources to applied research (43 percent of the total time budget), neglecting basic research (only 6 percent of the total time budget).

Table 3.28 Response patterns and response rate of research units

	Number of Registered Research Units 1997		Number of Responding Research Units 1997		Response Rate [a]
	no.	%	no.	%	%
Science Fields					
Architecture, Construction, Surveying	59	13.92	27	18.24	45.76
Biology, Chemistry, Medicine	77	18.16	31	20.95	40.26
Mathematics, Informatics, Physics	123	29.01	47	31.76	38.21
Electrotechnology, Mechanical Engineering	91	21.46	22	14.86	24.18
Economics, Social and Geosciences	74	17.45	21	14.19	28.38
Total	424	100.00	148	100.00	34.91

Note: a number of responding research units divided by total number of units multiplied by 100.

The funding of research projects depends heavily on public sources. Roughly 70 percent of the research budget comes from regional, national or international bodies. The national government is the most important source for research funding, the remaining 30 percent coming from business sources, mostly industry.

Contrary to this general picture, the field of electrotechnology and mechanical engineering obtains substantial external funding from industrial firms (56 percent of research funds). International institutions are not greatly involved in supporting research projects, due to the limited internationalisation of the research institutions, reflected for example in the limited participation in EU research programmes.

Table 3.29 Main activities and external funding sources of research units (1994-1996)

	Research Units					
	Architecture, Construction, Surveying	**Biology, Chemistry, Medicine**	**Mathematics, Informatics, Physics**	**Electro-technology, Mechanical Engineering**	**Economics, Social and Geosciences**	**Total**
Main Activities [a]						
Basic Research	16	19	29	6	24	21
Applied Research	32	34	29	43	28	33
Teaching	28	33	28	27	29	29
Transfer Tasks	17	9	11	13	11	11
Other	7	5	3	11	8	6
Funding Sources						
Business Funds	26	21	28	56	24	30
Public Funds from the Region	29	21	19	14	24	21
National Funds	23	36	38	16	33	31
Internat. Funds	16	20	13	7	18	15
Other [b]	6	2	2	7	1	3

Notes: **a** denotes percentage of total time budget of all scientific staff devoted to the activity;
b for example, funds from foundations.

B Networks and Network Formation

It is interesting to see that in the Barcelona innovation system research institutions do not appear to play a particularly active role in assisting innovation processes. They appear to be less frequently linked to manufacturing firms' R&D activities than vertical co-operation partners like customers or suppliers. Only 20 percent of manufacturing firms and 23 percent of the producer service firms reported that they co-operate with research institutions.

By contrast, the proportion of research institutions having co-operative relationships with other research institutions or businesses is quite high. Table 3.30 shows that 97 percent of the research institutions surveyed collaborate with other research institutes and 82 percent with businesses. In research collaboration with other research institutions, the fields of economics, social and geosciences, electrotechnology, mechanical engineering and biology, chemistry and medicine are the strongest. Through their embeddedness in national and international scientific networks, research institutions are potentially able to provide external knowledge to local businesses. The results of the survey underline that research institutions play a gateway function. In business collaboration, the fields of electrotechnology, mechanical engineering and biology, chemistry and medicine have the most intensive links with businesses.

Table 3.30 External co-operation partners of research units (1994-1996)

	Co-operation with Research Units			Co-operation with Businesses		
	a	b	c	a	b	c
Science Fields						
Architecture, Construction, Surveying	17	77	71	14	64	57
Biology, Chemistry, Medicine	18	82	61	17	77	41
Mathematics, Informatics, Physics	28	74	64	24	63	42
Electrotechnology, Mechanical Engineering	11	85	45	11	85	73
Economics, Social and Geosciences	13	87	62	8	53	25
Total	92	97	59	78	82	46

Notes: **a** denotes the number of co-operating research units.
b denotes the percentage of co-operating research units.
c denotes the percentage of research units reporting intensive or very intensive co-operation (i.e. indicating at least three regions with intensive contacts or two regions with very intensive contacts) out of the total in the given science field and region.

With respect to the intensity of research co-operation, Fig. 3.5 highlights the differences between the science fields. The highest co-operation intensity is found in architecture, construction, surveying, and electrotechnology and mechanical engineering. With the exception of this latter field, very intense external co-operation with research institutes is more frequent than with businesses (ranging between 71 percent for architecture, construction, surveying and 62 percent for economics, social and geosciences). Intensive links with businesses are reported only in the field of electrotechnology, mechanical engineering, with 73 percent of the institutions. The lowest intensity in co-operation with businesses is shown in economics, social and geosciences.

Where research institutions are involved in the innovation activities of firms, this is mainly in supporting their product innovations. In the Barcelona region, research institutions are particularly involved in the development of new products and the further development of products (see Table 3.31). The support of process innovation seems to be a less important task. As in other cases, it emerges that there are significant variations between the different science fields. Research units specialised in mathematics, informatics and physics mainly support firms in product innovation, an indication that they tend to be involved in the early stages of the product life cycle. On the other hand, research units operating in the fields of electrotechnology and mechanical engineering seem to concentrate on the further development of products, while those in the field of biology, chemistry and medicine are most concerned with process innovation. In the former case, the research units assist manufacturing firms in prolonging the product life cycle. In the latter case, research units support mass producing manufacturing firms in reducing production costs in order to maintain their competitiveness.

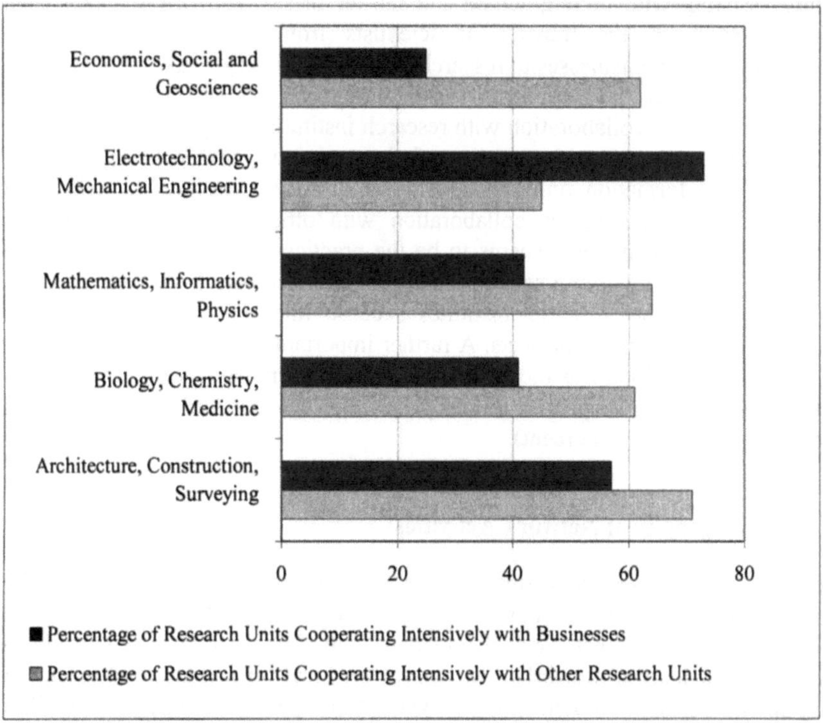

Fig. 3.5 Science fields active in external co-operation (1994-1996)

Research units tend to collaborate more with smaller firms, though less than might be expected. The sectoral structure of the manufacturing clients reflects the industry mix in the metropolitan region of Barcelona. The most important clients belong to the electrical engineering, energy and mining, chemical industry and steel, automotive and machinery branches. It emerges that the reasons for research or business collaboration differ significantly. 82 percent of research units surveyed indicate that the main reason for co-operation with businesses is to obtain practice-oriented impulses for research projects. Financial motives like reducing dependence on public contracts (63 percent) or the realisation of costly research projects (58 percent) follow a long way behind.

Table 3.31 shows the results of the question put to research units about the kind of support they offer to firms. The most important activity was the joint supervision of undergraduate and master theses, followed by testing services and consultancy. Besides this overall pattern, each science field appears to have its own particular specialisation. The field of architecture, construction and surveying is mostly involved in consultancy, while biology, chemistry and medicine focus on conducting tests and measurements. Electrotechnology, mechanical engineering and economics, social and geosciences show a similar specialisation, mainly assisting with undergraduate and master theses, and offering personnel training. However, the transfer of scientists from research institutions to businesses, or from businesses to research institutions is very weakly developed.

With regard to the support of innovation activities in firms during the innovation process, collaboration with research institutions appears to concentrate on the early, pre-competitive stages of the innovation process. The general exchange of information and the generation of new ideas are the two most important phases. As far as collaboration with other research institutions is concerned, the driving force seems to be the practice of sponsoring agencies to fund joint projects involving several institutions. 62 percent of research institutes co-operate with other research institutes because financial sponsorship is only available for collaborative projects. A further important motive which emerged in the survey of Barcelona was the tendency of less well known research institutes to seek to raise their own profile or image through co-operation with prestigious research institutions (42 percent).

C The Geography of Network Activities

The activities of research institutions can have a considerable impacts on the regional economy when the businesses whose innovation efforts supported are mainly local. On the other hand, if research institutions participate principally in international scientific networks, the impact on local businesses is likely to be more in the form of knowledge spillovers than direct economic effects. This suggests that the geographical distribution of co-operation partners will vary depending on whether they are businesses or other research institutions. Indeed, this is confirmed by the results obtained from the survey.

Fig. 3.6 illustrates the importance of the metropolitan region of Barcelona as a location for co-operation partners, both for businesses and other research institutions. More than 60 percent of the co-operation partners of surveyed firms are located there. Table 3.32 gives a detailed breakdown of the location of business partners. In general, the intensity of co-operative activity declines with increasing distance. Next to the metropolitan level, the Catalan and national level are important locations of businesses. The support of businesses outside Spain is relatively unimportant. While this geographical pattern is true for nearly all fields, biology, chemistry, and medicine turned out to be rather more internationally orientated, with 42 percent of the research institutes surveyed supporting innovation activities in businesses outside the European Union.

Table 3.31 Forms of co-operation offered by research units to business (1994-1996)

	Services or Assistance Offered to Manufacturing Firms by Science Fields					
	Architecture, Construction, Surveying	**Biology, Chemistry, Medicine**	**Mathematics, Informatics, Physics**	**Electro-technology, Mechanical Engineering**	**Economics, Social and Geosciences**	**Total**
***Form of Co-operation* (as percentages)**						
Testing Services	42	68	44	60	23	50
Supply of Equipment	42	27	33	45	8	33
Consultancy	63	36	52	60	62	53
Undergraduate/Master Work (theses)	47	41	67	60	69	56
Doctoral Work (theses)	37	45	48	40	62	46
Training of Business Personnel	47	45	44	60	62	50
Personnel Transfer from Businesses	11	27	11	30	23	20
Personnel Transfer to Businesses	21	27	30	35	31	29

The location of research institution partners follows a different pattern. Unlike the geographical distribution of businesses, co-operation with other research institutes is highest outside the European Union (79 percent of the research units surveyed). However, the national and European levels remain important as locations for scientific co-operation partners. These results demonstrate that research institutes are embedded in national and international scientific networks and potentially able to provide external knowledge to local businesses.

The importance of spatial proximity is also underlined in Table 3.36. In the survey, research institutions were asked to state whether spatial proximity was perceived as an important factor for co-operation with businesses or not. Only 18 percent of the research units reported that spatial proximity was not a significant factor. Extremely low negative responses were found in the fields of electrotechnology, mechanical engineering and biology, chemistry and medicine. With respect to the different innovation phases, it seems that spatial proximity matters in more advanced phases of the innovation process, i.e. during conception/front-end development, prototype development and pilot applications (e.g. electrotechnology, mechanical engineering).

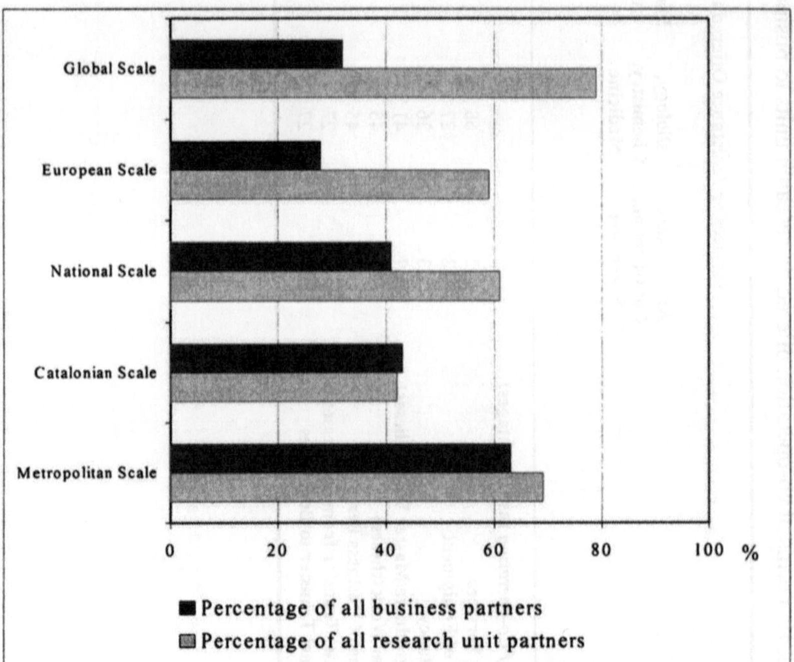

Fig. 3.6 Location of co-operation partners of research units with intensive and very intensive co-operation relationships (1994-1996)

Table 3.32 Location of business partners, differentiated by science field of the research units (1994-1996)

	Research Units by Science Field [a]					
	Architecture, Construction, Surveying	**Biology, Chemistry, Medicine**	**Mathematics, Informatics, Physics**	**Electro-technology, Mechanical Engineering**	**Economics Social and Geo-sciences**	**Total**
Metrop. Region	59	68	55	86	55	63
Catalonia	22	61	40	68	23	43
Spain	41	35	34	73	32	41
European Union	19	29	32	36	23	28
Rest of World	19	42	34	32	32	32

Note: a percentage of research units co-operating with businesses by science field and geographical area

3.6 The Barcelonese Way of Networking

The descriptive analysis of the innovation activities of manufacturing firms, producer service firms and research institutions shows that the R&D activities of these major actors in the metropolitan innovation system of Barcelona are closely interrelated. In this section, we turn our attention an investigation of the extent to which networking is influenced by firm-specific attributes, as suggested by the resource-based view of the firm (see Fischer and Varga 2000). This research question has been analysed via logit modelling in an attempt to overcome the difficulties inherent in bivariate analysis by means of rigorous multiple regression modelling for data with a dichotomous response variable (for more details see Fischer and Nijkamp 1985).

The restricted nature of the postal survey limited the number of independent variables available. However, as basic profile attributes, it was possible to incorporate the following variables: organisational structure, turnover, employment size as a proxy for scale economies, years in operation as a proxy for a learning-by-doing effect, ownership and export intensity. There were also proxies for technological resources and opportunities, such as R&D expenditure, in-house research skills, presence of on-site R&D facilities, technological opportunities and innovation competence as defined in Table 3.33.

In the results given below, with the exception of the continuous variables, the parameter estimates may be interpreted with respect to the reference category. This is a function of the particular parametrisation used in estimating the model and is set to zero. The reference category consists of domestic independent establishments in the low technology sector with no on-site R&D facilities and a low level of export orientation.

Table 3.33 Firm-specific variables included in the logit analysis

Independent Variables	Variable Type	Variable Definition
Basic Profile Attributes		
Organisational Structure	dummy	=1 denotes multi-unit =0 otherwise
Turnover	continuous	Annual turnover (averaged over 1994-1996)
Employment Size	continuous	Total employment per establishment (natural logarithm)
Years in Operation	continuous	Establishment age, calculated as '1998 minus years formed'
Ownership	dummy	=1 denotes foreign (some share of total capital) =0 otherwise
Export Intensity	dummy	=1 denotes high export intensity (over 50% of turnover) =0 otherwise
External Co-operation	dummy	=1 denotes the presence of any external link =0 otherwise
Proxies for Technological Resources and Opportunities		
R&D Expenditure	continuous	Annual R&D expenditure in % of turnover (averaged over 1994-1996)
In-House Research Skills	continuous	Research personnel as % of R&D personnel (averaged over 1994-1996)
Presence of On-Site R&D Facility	dummy	=1 denotes presence =0 otherwise
Technological Opportunities	dummy	=1 denotes high technology sector (using the definition of Hatzichronoglou 1997) =0 otherwise
Innovation Competence	continuous	Share of turnover accounted for by new or improved products (averaged over 1994-1996)

Table 3.34 indicates the degree to which firm-specific attributes increase or decrease the probability (strictly, the logarithmic odds) of external networking. There is no claim that the results presented in the table should in any sense represent an 'optimal' model. Rather, the approach is essentially exploratory and the intention is to demonstrate which variables are important and to identify their direction. The study relies on a subset of 95 manufacturing firms which provided the necessary information.

Table 3.34 Network activities of manufacturing firms: Parameter estimates (t-values in brackets[a])

Variable	(a) Customer Networks	(b) Manufacturing Supplier Networks	(c) Producer Service Provider Networks	(d) Producer Networks	(e) Science-Industry Relations
Constant	2.33	-1.83	-2.73	-0.49	-4.74
	(1.44)	(-1.04)	(-1.32)	(-0.25)	(-2.25)
Organisational Structure	0.80	0.12	-1.72	1.85	1.56
	(1.24)	(0.20)	-(2.42)	(2.62)	(2.28)
Turnover	0.00	0.00	0.00	0.00	0.00
	(1.15)	(0.38)	(-0.78)	(-0.93)	(-0.66)
Employment (Log)	-1.38	1.40	3.15	-0.29	0.79
	(-1.43)	(1.34)	(2.40)	(-0.24)	(0.66)
Years in Operation	-0.01	-0.02	-0.01	-0.03	0.01
	(-1.41)	(-2.10)	(-1.19)	(-1.61)	(1.01)
Ownership	0.95	0.26	-0.10	-0.37	-0.30
	(1.61)	(0.47)	(-0.16)	(-0.50)	(-0.47)
Export Intensity	-0.04	0.58	-0.70	1.63	1.36
	(-0.05)	(0.80)	(-0.91)	(1.97)	(1.82)
R&D Expenditures	0.04	0.04	0.10	-0.01	-0.03
	(0.97)	(0.97)	(1.56)	(-0.34)	(-0.86)
In-House Research Skills	-1.26	-1.15	-1.53	-0.66	0.58
	(-1.50)	(-1.41)	(-1.69)	(-0.64)	(0.60)
On-Site R&D Facility	-0.34	-0.11	-0.89	-0.21	0.65
	(-0.57)	(-0.19)	(-1.24)	(-0.30)	(0.98)
Technological Opportunities	0.28	-0.03	-0.94	0.83	1.59
	(0.53)	(0.05)	(-1.64)	(1.37)	(2.61)
Innovation Competence	0.02	0.01	0.01	0.00	0.01
	(1.98)	(1.28)	(0.84)	(-0.32)	(0.81)
Number of Observations	95	96	95	95	95
Log-Likelihood	-55.11	-57.78	-49.78	-41.93	-45.46
Likelihood Ratio Test	19.69	15.47	17.34	16.51	24.28
Rho Squared	0.16	0.13	0.24	0.36	0.31
Adjusted Rho Squared	0.08	0.05	0.16	0.28	0.23
Prediction Success (%)	68.00	62.00	77.00	80.00	78.00

Note: a significant at the 10 % level at least.

Columns (a) and (b) of Table 3.34 indicate that there is very little variability in the case of both customers and manufacturing supplier relationships. A very low level of adjusted rho-squared is accompanied by a predictive success of 68 percent and 62 percent respectively. Only innovation competence (in the case of customer networks) and years in operation (in the case of manufacturing supplier networks) are significant, but they have only a weak effect. In the case of producer service provider networks, it should be clear from Table 3.34 (c) that employment size and organisational structure are the dominant variables, reflecting the fact that the probability of networking is much higher in multi-unit than in independent establishments. Moreover, the larger the establishment, the higher the probability of networking with producer service providers.

Columns (d) and (e) of Table 3.34 show that organisational structure is also a significant factor in the cases of producer networks and science-industry relations. While high export intensity plays an important role in the case of producer networks, technological opportunities matter in the case of science-industry relations. This result suggests that firms exhibiting a higher degree of technological opportunities co-operate more with research institutions.

3.7 Concluding Remarks

In this chapter, we have looked in some detail at the four key building blocks of the metropolitan innovation system of Barcelona: the institutional setting, the manufacturing sector, the sector of producer services, and the science and research sector. The analysis has shown clearly the direct impact of institutional transformations since Spain's entry into the European Community. Not only has the country had to open its market to foreign competition, but the institutional setting too has changed as a consequence of the decentralisation process and the consequent strengthening of the regional level. One of the major tasks in economic policy is to overcome the weaknesses of the R&D system inherited from the Franco era.

Seen in an international perspective, the Spanish innovation system is lagging behind the leading nations in Europe, the US and Japan. Due to the increased competitiveness of East European Countries and Southeast Asia, the Spanish economy faces a special challenge. Producer service firms and research institutions, as well as manufacturing firms, being the main actors in the innovation system, will have to increase their innovative capacity and capabilities in order to be able to successfully continue the catching up process initiated in 1986. Still today, there is a strong dependence on foreign technology in the most important branches like electronics, transport equipment, chemical industry.

At a regional level, the Spanish innovation system is dominated by two metropolitan regions: that of the capital, Madrid, with the highest concentration of public-funded R&D initiatives, and Barcelona, with the highest level of private involvement in R&D. In terms of firm-related figures, such as industrial R&D expenditure or patents applied for, the metropolitan region of Barcelona is ahead of

Madrid. An important reason for Barcelona's greater innovativeness is the presence of multinational corporations. However, although these multinationals introduce technology-intensive products and innovative production processes, they remain rather isolated. Apart from this modern and internationally competitive segment, the manufacturing sector in the metropolitan region of Barcelona is characterised by small and medium-sized firms with low levels of R&D.

Barcelona's leading role within the national innovation system in terms of business R&D is also a consequence of political efforts at the regional level. Based on wide consensus, which is expressed in a strong social and cultural identity, the Catalan and local government have put great emphasis on the internationalisation of the innovation system. The regional government is taking an active role in transforming the institutional set-up and the regional plan for research is only one example of these efforts.

Empirical evidence from the surveys of manufacturing firms, producer service providers and research institutions in the metropolitan region of Barcelona illustrate the changes taking place. The catching-up process of the regional economy is visible in the innovation efforts of the firms surveyed. From being a long way behind, Barcelona's firms are becoming increasingly innovative although, in comparison to Stockholm for example, the innovations tend to be incremental rather than radical. It seems that during the catching up process, the actors of the metropolitan innovations system of Barcelona first have to develop complex knowledge bases through learning-by-doing and using in order to bring out radical inventions in the future, as in the case of more mature, knowledge intensive regional economies.

With respect to interrelations between the main actors in the innovation system, the most important results can be summarised as follows. *First*, a firm's own R&D efforts and the accumulation of production experience are the most important sources of innovation. External co-operation is relevant and more important than the acquisition of licenses. In-house R&D facilities and external co-operation have a positive and very significant impact on the metropolitan innovation system of Barcelona. Nearly 80 percent of the manufacturing firms are involved in external co-operation networks. The network propensity varies between industrial sectors and firm size. The electrical and optical equipment and chemicals, plastic and rubber sectors and medium-sized firms are those most active in networking.

Second, the co-operation activities of manufacturing firms is based much more on vertical relationships (customers, suppliers) than on horizontal relationships (producer networks and science-industry relations). Co-operation with research institutions is of minor importance and more likely in single firms than in multi-establishment organisations or high-tech firms.

Third, the geographic distribution of co-operation partners is concentrated in the metropolitan region itself. In general, firms tend to prefer local and national co-operation partners rather than international ones. Innovative firms focus more strongly on external co-operation than non-innovative firms. The co-operation partners of the innovative firms are much more concentrated in the metropolitan region of Barcelona than the non-innovative ones. This might indicate the importance of the tacit nature of the knowledge concerned.

Fourth, the most innovative producer service firms belong to the computer software sector. Technical consultancy and market research, and larger producer service firms, are very strongly orientated towards manufacturing firms. The geographical distribution of manufacturing clients varies between the sectors. The more business-orientated producer services tend to have manufacturing clients exclusively in the metropolitan region of Barcelona, whereas more technically orientated producer services also have clients at the national and European level.

Fifth, the role of research institutions in the metropolitan region of Barcelona is still weak. For manufacturing firms and producer services, nearby research institutions are not the main source of innovation. On the other hand, most of the research institutions have co-operative relationships with other research institutes or businesses. Whereas all the science fields are involved in co-operative projects with other research institutions, business collaboration is significant only in electrotechnology, mechanical engineering and biology, chemistry and medicine. Research institutions are essential in training the potential workforce. The joint supervision of undergraduate and master theses is the most important activity offered to businesses, followed by testing services and consultancy. While the business co-operation partners are concentrated in the metropolitan region of Barcelona, the co-operation partners in research may also be found at the national and international level. Spatial proximity is a very important factor for research institutions in electrotechnology, mechanical engineering, and biology, chemistry and medicine, during the more advanced phases of the innovation process.

The results obtained illustrate the weaknesses of the surveyed firms from the innovation point of view. Manufacturing firms in particular need to devote far more effort to R&D activities in order to remain competitive. This is particularly crucial for the small and medium-sized firms which have a low innovative performance. Policy therefore needs to focus on strengthening the R&D capabilities of the smaller firms. Only in this way can the necessary absorption capacity for using external knowledge be established. In addition, the low level of internationalisation, visible both in terms of market reach and the geographical scope of the innovative network partners, is a matter of concern.

4 The Stockholm Metropolitan System of Innovation

The metropolitan region of Stockholm has essentially a polycentric structure in which the city of Stockholm itself, the Swedish capital, is by far the strongest player. The rest of the Mälar region can be said to constitute the hinterland of metropolitan Stockholm (see Fig. 4.1). The ample availability of land in Sweden implies that the pressure is not as large on scarce land resources for urban development as in other European metropolitan regions. Lake Mälaren is an environmental asset and a trade infrastructure, at the same time being the major supplier of drinking water to all the urban settlements in the area. For the moment there is no comprehensive physical plan that controls the future land-use in the region, which houses a number of small and medium-sized urban settlements. The hinterland of the Stockholm region used to be the core of the Swedish manufacturing industry.

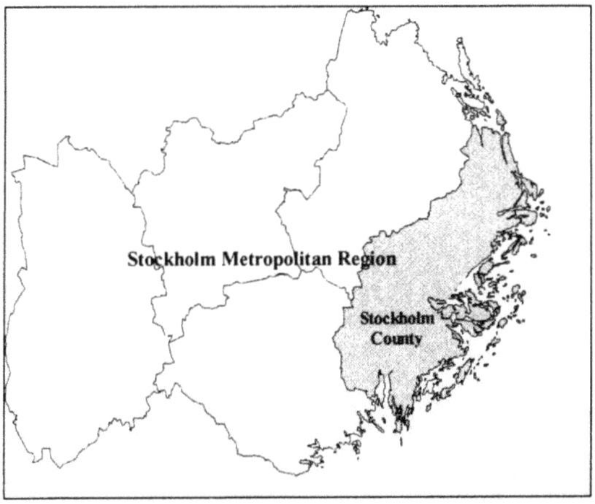

Fig. 4.1 The geographical and administrative structure of the metropolitan region of Stockholm

In principle, it could be argued that the resources in the region are enough to house a substantially larger share of Sweden's population than the current 20 percent or so. One Swedish view is that this is neither likely from an economic point of view nor desirable from an environmental standpoint. Much work is currently going into rethinking the value structure of land-use planning, upgrading the importance of the landscape and its associated ecological systems. Some pressure in the Mälar region comes from the need to expand airport capacity and other transport infrastructure systems. In comparison to the pressures exerted on the scarce land resources in other parts of Europe, these problems might seem less difficult to handle.

The extended Stockholm region is the industrial and political heartland of Sweden. It houses the capital city of Stockholm with its collection of public goods and nationally-oriented public sector activities. Some of the strongest manufacturing corporations in Sweden have their home base for innovations and R&D in this part of Sweden. Together with the Copenhagen-Malmö region, the Mälar region is a development core of the IT and knowledge society at the European level as well as globally. The region has a diversified and well-functioning transport and communications infrastructure.

Fig. 4.2 shows that the metropolitan regions of Barcelona, Stockholm and Vienna occupy positions in the upper part of the size distribution of the 250 largest functional urban regions in Europe. The urban system's uppermost part contains regions with a somewhat smaller population number than the straight-line prediction of Zipf's rule applied to Europe as a whole. The figure also reveals that the smallest regions do not fully belong to the integrated European urban system.

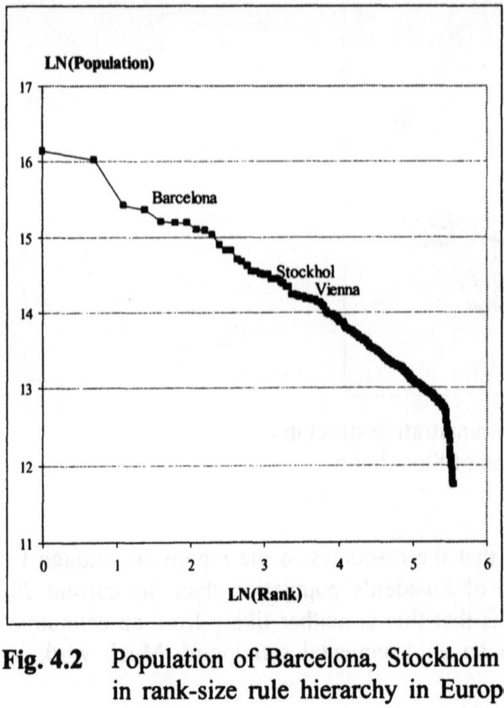

Fig. 4.2 Population of Barcelona, Stockholm and Vienna metropolitan regions in rank-size rule hierarchy in Europe at beginning of the 1990s (log scales)

4.1 Metropolitan Region With a Service Specialisation

The Stockholm metropolitan area is the most service-specialised functional urban region in western Europe, with more than three out of four jobs in the private and public service sector (see Fig. 4.3). Although one of the largest urban regions in Europe, Barcelona still has a strong concentration of manufacturing activity.

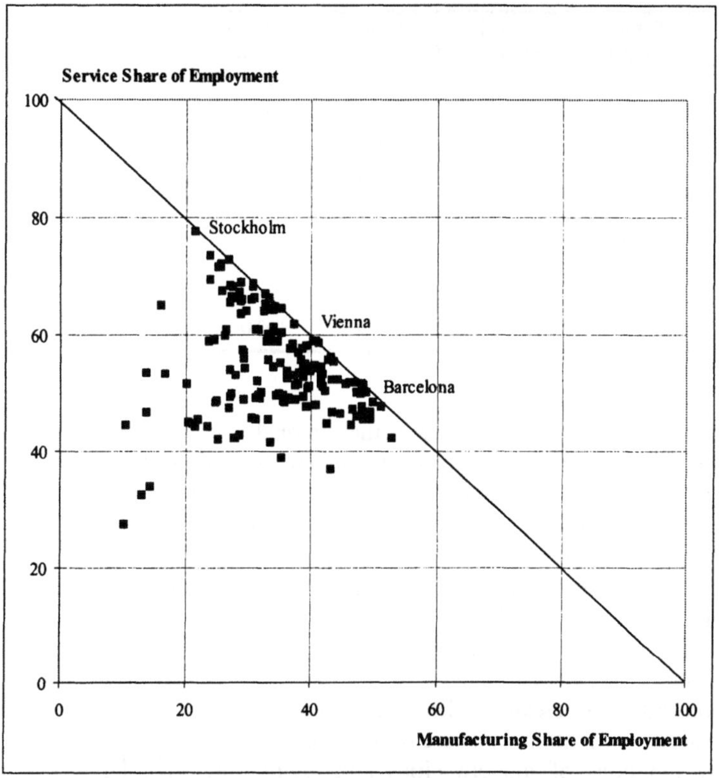

Fig. 4.3 The specialisation pattern of some European metropolitan regions at the beginning of the 1990s

Sweden has a three-tier government structure. The municipalities form the local government level. There are some 280 in total, with an average population size of just over 30,000 inhabitants, see also Table 4.1. They have the authority to set both income and property taxes. During recent years, a number of tasks in public service provision have been decentralised to the municipalities, which have consequently been given a greater degree of freedom in organising their activities through alterations in the municipal judicial system. As a result of political action, the role of the municipalities has been concentrated to service management rather than service production. The responsibilities of the municipalities extend over a broad range of public sector activities from child-care to environmental management and land-use planning. In recent years, they have also extended their activities in the field of local educational and industrial policy.

Table 4.1 The administrative structure of middle Sweden and per capita income of the constituent regions in 1995

County/Region	Population (1,000s)	Local Authorities	Municipal Size (1,000 persons)	Income/ Capita
Stockholm	1,642	25	65.7	121.2
Uppsala	269	6	44.8	97.5
Södermanland	256	7	36.6	96.0
Västmanland	258	10	25.8	98.5
Örebro	273	11	24.8	96.1
Metropolitan Region	2,698	59	45.7	111.7
Sweden	8,591	278	30.9	100.0

The Swedish system for policy-making at the regional level is currently under revision. Experiments are being made to strengthen the responsibilities of the regional level by encouraging municipalities to work together, and also introducing regional parliaments, as in the Skåne region in southernmost Sweden. The aim is to place responsibility with the appropriate regional level to manage the competence needed for decision support and to increase economic efficiency in public service production. In Sweden, this development is being promoted in parallel to an increased involvement of private initiatives in local and regional service provision. The Swedish model is for the public sector to retain the financing responsibility but create markets for private firms to bid for contracts in the health-care sector, education, and care of the elderly.

The gateway functions of modern cities and regions are to a large extent connected to the infrastructure systems, see Andersson and Andersson (2000). Until recently, industrial and regional policies in Sweden were managed separately, but economic growth, transport infrastructure provision and regional development have now been brought under one joint Ministry. The planning of large infrastructure projects is organised more traditionally than in other European countries. The primary initiative still lies to a large degree with the transport planning agencies. One result has been long and conflict-prone decision processes, especially in the metropolitan regions, see Hall, Prud'homme and Snickars (1994). During the late 1990s, though the pace of economic and population growth increased, very little housing construction took place in the Stockholm region. The picture which emerges is that of a region which is one of the leaders in R&D-led economic growth, with focal points in the information technologies (IT) and pharmaceutical sectors.

4.2 Some Features of the Swedish National Innovation System

R&D effort in the Swedish business sector grew rapidly during the 1990s. Table 4.2 indicates the level of commitment in some manufacturing sectors in the first part of the decade. R&D in person-years increased by 18 percent 1993-95. The increase was even larger in terms of R&D expenditure, which rose by 22 percent in fixed prices during the same period (see NUTEK 1997). The corresponding increase in R&D spending in manufacturing was 21 percent. Dominant industries in manufacturing R&D are the transport equipment and telecommunications sectors (part of 'other manufacturing'). Other sectors directing a large proportion of their value-added into R&D are machinery, pharmaceuticals and instruments. In 1995 these accounted for 85 percent of total R&D expenditure in the business sector.

Table 4.2 Some features of Sweden's R&D system in 1995

Industry	Person-Years of R&D		R&D Expenditure (Million Euro)		R&D Purchases (Million Euro)	
	1995	Annual % 1993-95	1995	Annual % 1993-95	1995	Annual % 1993-95
Transport Equipment	8,714	6.4	1,108	19.6	137	54.0
Electronic Equipment	6,610	7.3	857	7.4	619	87.6
Pharmaceuticals	4,026	16.4	717	18.4	127	45.5
Other Manufacturing	14,938	5.1	1,629	13.8	131	58.5
Non-Manufacturing Industries	7,140	17.1	793	14.8	326	18.6
All Manufacturing Industries	34,288	7.0	4,312	14.5	1,013	71.2
Total Business Sector	41,636	8.6	5,120	14.7	1,343	52.6

R&D expenditures outside manufacturing have increased considerably in recent years. In fact, the most rapid increase in R&D resources has been in non-manufacturing industries, with particular emphasis on R&D firms, R&D institutions and computer consultancy. There was also a sharp increase in the amount of R&D services bought by the manufacturing sector. A large proportion of these purchases involved producer service industries. The telecommunications industry was particularly prone to buy R&D from service firms.

Sweden performs about 1.5 percent of the R&D in the world. As Fig. 4.4 shows, the share is above the average in medicine, biology, and physics and mathematics. The intensity of scientific work measured in terms of the number of scientific publications per capita is among the highest in the world. This is clearly evident in Fig. 4.5, which reveals that researchers in Sweden, together with Switzerland and

Denmark, had more than 3,000 articles published per 100,000 inhabitants in the period 1991-95.

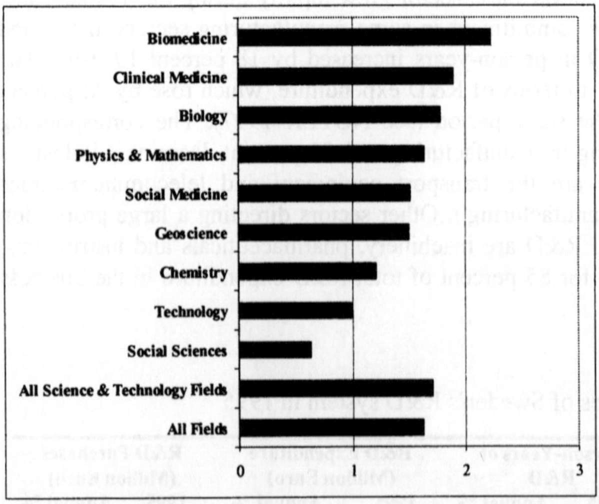

Fig. 4.4 Sweden's world share of R&D in 1997 (percent)

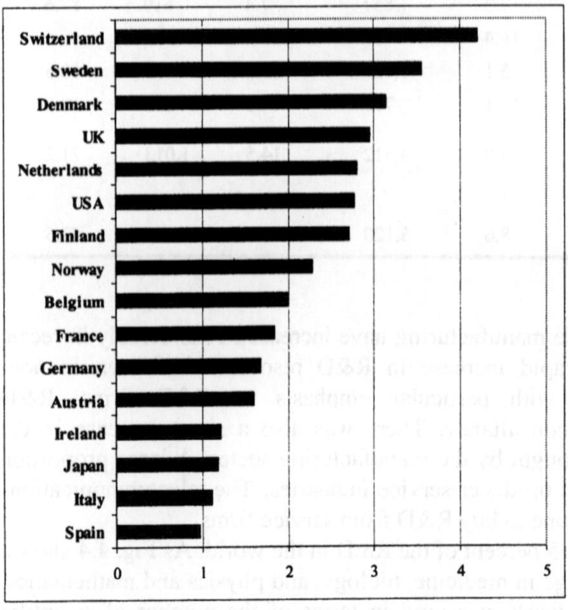

Fig. 4.5 Number of scientific publications per 100 inhabitants (1991-1995)

When R&D specialisation is measured in economic terms, it turns out that Sweden spent a larger share of GNP on R&D in both the public and private sectors than any other country. In 1995 the share invested was 3.6 percent of GNP, substantially higher than countries as Japan, Korea and USA, see Fig. 4.6.

If we compare internationally across sectors, we find that Sweden's advantage, compared to the average for the EU plus Japan and the US, is highest in the non-manufacturing sector (see Fig. 4.7). However, the manufacturing sectors where Sweden has strong and internationally competitive firms – telecommunications, vehicles, pharmaceuticals, electrical machinery – also contribute to Sweden's strong R&D profile. Fig. 4.8 completes the picture by providing an illustration of the international role of Sweden in the R&D efforts in the higher education system. The figure shows that, also in this regard, Sweden has a leading position, spending a larger share of GNP on R&D than any other country. The Swedish specialisation in R&D is stronger than in Japan, and the gap between Sweden's spending and the level of effort in other European countries is considerable.

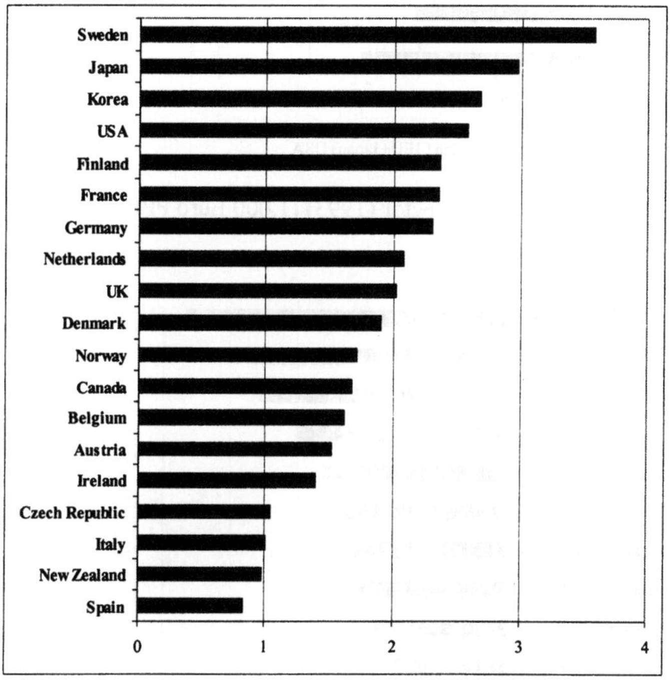

Fig. 4.6 Gross national R&D expenditure by country in 1995 (percent of GDP)

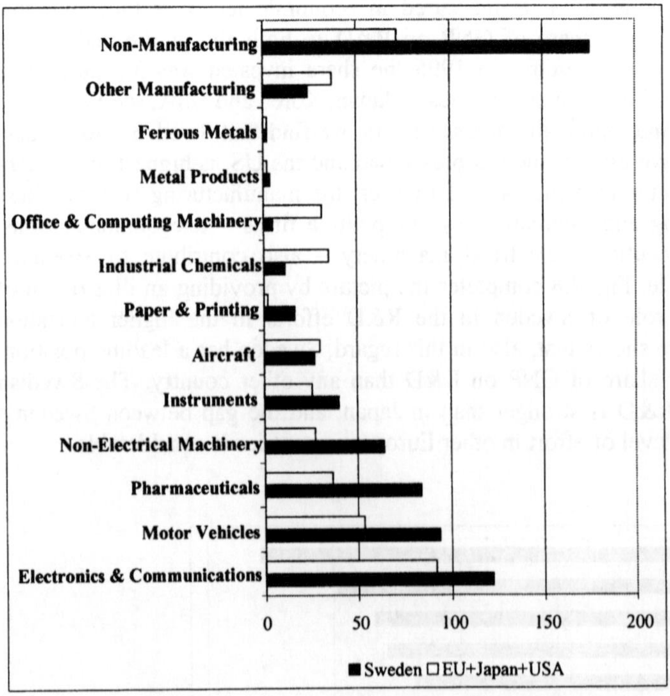

Fig. 4.7 R&D expenditure in business sector (1993) (1,000 Euro PPP per capita)

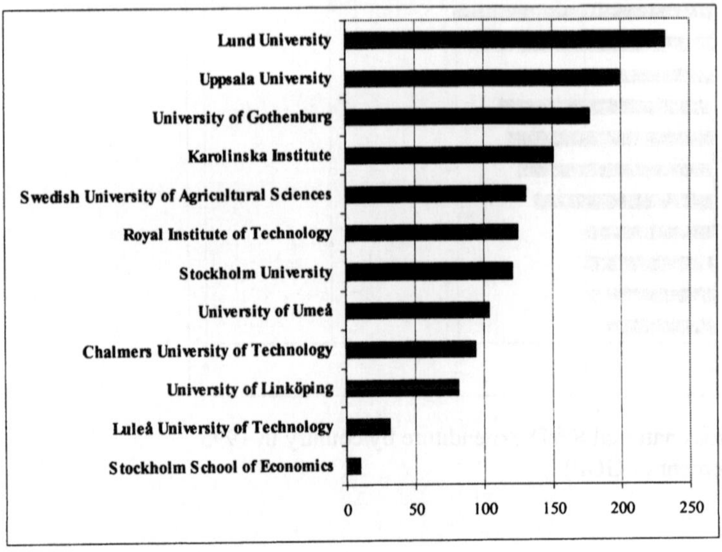

Fig. 4.8 R&D at universities and regional colleges in Sweden (1995-1996) (in million Euro)

In the picture of gross domestic expenditure on R&D provided in Fig. 4.6, we can see that Sweden spends more than twice as much as Austria and more than four times Spain. It is of course not straightforward to compare expenditure levels across countries with quite different educational and R&D systems, both at the basic level and at the university level. In the table, the Austrian figures are missing, but it is estimated that the level is approximately the same as Germany. Even so, these national level considerations are important to keep in mind when the innovation systems of the three metropolitan regions of Barcelona, Stockholm and Vienna are compared.

4.3 The Institutional Set-Up

The R&D sector is an important element in the Swedish innovation system. During recent years this has been politically recognised, and institutional changes made in the R&D financing system, with a special funding agency being established for the support of the Swedish innovation system. Many of the resources will go to research departments within the university system and to the public, semi-private and private research institutes that engage in R&D related to innovation outside the universities. The regional structure of the Swedish R&D and innovation system will determine to what extent the promotion of innovation activities will affect the regional balance. As shown in Fig. 4.8 and Fig. 4.9 the metropolitan regions, and particularly the Stockholm region, has a very strong role in the R&D system. One may say that the Stockholm metropolitan region, and its hinterland around Lake Mälaren, are the regions where R&D and innovation is most strongly concentrated in a country that is already one of the most R&D and innovation-oriented in the world. This should be reflected in the innovation potential of the region and the way the innovation networks operate.

The figures show that Sweden's largest university is located in Lund. The university capacity in the metropolitan region is made up of Uppsala University, the Karolinska Institute with its famous medical school, Stockholm University, the Royal Institute of Technology, and the Stockholm School of Economics. Besides these big actors there are a number of other institutions for higher education and research in the humanities in the Stockholm region.

It should be mentioned that the Swedish Agricultural University has an important centre in Uppsala, as well as being spread over several other locations in the country. During recent years, Södertörn University College has been established, further increasing the research facilities in the southern part of the Stockholm region. In spite of this, it is generally agreed that the potential for further expansion of the higher education and public R&D systems in the Stockholm region is substantial. A clear indication of the potential is the fact that many of the graduates from universities and university colleges across the country find their jobs in the Stockholm region.

Name	Main Sponsor	Research Area	Main Location	Turnover (mill. Euros)
STFI	NUTEK	Paper and Pulp	Stockholm	18.2
IVF	NUTEK	Production Engineering	Gothenburg	15.4
IVL	SNV	Environment	Stockholm	9.4
Skogforsk	SJFR	Forestry	Uppsala	8.3
Trätek	NUTEK	Wood	Stockholm	8.1
SICS	NUTEK	Computer Science	Stockholm	5.8
SIK	NUTEK	Food	Gothenburg	5.6
IMC	KTH	Microelectronics	Stockholm	5.5
IM	NUTEK	Metals	Stockholm	5.2
MEFOS	NUTEK	Metallurgy	Luleå	5.0
IFP	NUTEK	Fibres and Polymers	Gothenburg	4.5
YKI	NUTEK	Surface Chemistry	Stockholm	4.3
Packforsk	NUTEK	Packaging	Stockholm	3.3
SISU	NUTEK	Information Systems	Stockholm	3.2
KI	NUTEK	Corrosion	Stockholm	2.9
IMT	NUTEK	Media Technology	Stockholm	2.9
JTI	SJFR	Agricultural Engineering	Uppsala	2.6
CBI	BFR	Cement and Concrete	Stockholm	2.5

Notes: NUTEK = Swedish Board for Industrial and Technical Development, SJFR = Research Council for Forestry and Agriculture, BFR = Swedish Council for Building Research, SNV = National Swedish Environmental Protection Agency, KTH = Royal Institute of Technology.

Fig. 4.9 Some major industrial research institutes in Sweden (1995)

Fig. 4.9 shows the geographical location and industrial orientation of Sweden's largest research institutions. Fourteen of the 18 institutes listed are located in the Mälar region, where the financing agencies are also located. This picture illustrates the strongly dominating role of the Stockholm region in the Swedish R&D system. The question is what this implies for the innovation networks in the region and to what extent it builds up the innovation potential.

4.4 The Manufacturing Sector

A The Structure of Manufacturing in the Region

In this section we present the results of the questionnaire responses from 450 manufacturing companies in the Mälar region. Some 190 of these are to be found in the Stockholm region itself, while the remaining ones are spread across the four other counties that comprise the extended functional economic region making up the Stockholm metropolitan area.

Almost 65 percent of the gainfully employed and 60 percent of the total population in the whole metropolitan region are found in the county of Stockholm. The study concludes that the Mälar region is not an economically homogenous region but one that differs with respect to manufacturing employment. In this respect, it is mainly Stockholm that deviates from the other sub-regions. The share of labour and capital intensive industry in Stockholm is considerably lower than in the rest of the region, and the share of R&D intensive industry is higher, as shown in Table 4.3.

How is the difference between the firms in the Stockholm region and the rest of the metropolitan region in the industrial structure? We can observe that the share of employment in the labour and capital-intensive industry is lower in the Stockholm region and considerably higher in the R&D intensive sector. There is, thus, a concentration of protected industry in the rest of the metropolitan region to the same extent that the knowledge-intensive one is higher in the core of the region. The R&D intensive sector is more important in the Stockholm region than in the rest of the region. Large companies are more prevalent outside of the Stockholm region. Those companies that spend more than 10 percent of the turnover on innovation activity are more frequent in number in the Stockholm region.

Table 4.3 Distribution of manufacturing employment in the metropolitan region (number of responses in brackets)

Manufacturing Sector	Stockholm County (Percent)	Rest of Region (Percent)	Metropolitan Region (Percent)
R&D-Intensive	22.0	11.6	16.0
Knowledge Intensive	17.3	30.1	24.7
Capital Intensive	5.2	8.5	7.1
Labour Intensive	21.5	27.8	25.1
Protected	34.0	22.0	27.1
Total	(191) 100	(259) 100	(450) 100

A majority (63 percent) of the responding firms in Stockholm County had fewer than 50 employees, see Table 4.4. This is a much lower share compared to 1993, when 97 percent of a total of 68,500 were classified as small firms. Approximately three out of four firms both in Stockholm County and the rest of the metropolitan region stated that they perform R&D in their establishments in the region.

Table 4.4 Distribution of firms by size classes

Employment Size	Stockholm County (Percent)	Rest of Region (Percent)	Metropolitan Region (Percent)
0-49	75.5	63.2	68.5
50 or More	24.3	36.8	31.5
Total	(189) 100	(258) 100	(447) 100

In Table 4.5 we see that most firms in Stockholm County spent less than 10 percent of their turnover on innovation. Innovation costs are conceived here as expenditure on R&D, construction and design. Licenses and external purchases are also counted. One explanation could be that due to their geographical proximity, the firms in this part of the metropolitan region have closer connections with one another and also the R&D agencies located in the Stockholm region.

Table 4.5 Distribution of innovation costs as share of turnover

Share of Turnover	Stockholm County (Percent)	Rest of Region (Percent)	Metropolitan Region (Percent)
0-4 Percent	56.5	66.0	62.0
5-9 Percent	20.3	21.3	20.9
Over 10 Percent	23.2	12.8	17.2
Total	(138) 100	(188) 100	(326) 100

Half of the surveyed firms in the Stockholm region are exclusively manufacturing units. The other half produces both goods and services. The situation is different for the rest of the metropolitan region where manufacturing firms dominate. If the firms are divided according to dominant product group, it turns out that in the Stockholm region systems products and consumer goods dominate, making up a quarter of the turnover each. In the rest of the metropolitan region, the biggest product groups are semi-finished products and investment goods having 23 and 26 percent of the turnover, respectively. The share of foreign-owned firms is somewhat higher (24 percent) in the rest of the region than in the Stockholm region (20 percent). A considerable share of the output from manufacturing industry is exported. In the capital, knowledge, and R&D intensive industries, the export share is above 30 percent, as shown in Fig. 4.10.

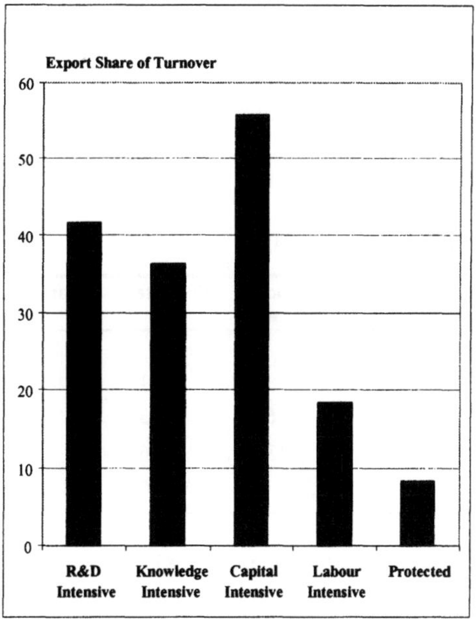

Fig. 4.10 Export shares for manufacturing firms in the metropolitan region

B Innovation Activities of Manufacturing Firms

Innovations are important for the development of a region. For the firms in industrialised nations to achieve continued growth, they must be able to present on the market new and enhanced products and processes. The competition nowadays is very tough, and innovations can be seen as a way of withstanding the high pressure from both near and distant markets. In this context, we focus on the role of regional innovation systems in terms of enhancing innovation potential. Three indicators have been used to give an overview of the scope of innovation activity at the level of the firm, the R&D costs, construction and product design, the number of product and process innovations, and the number of firms having filed patent applications.

The highest share of firms spending more than 10 percent of their turnover on R&D, construction and design is in the R&D-intensive industries. Among these, product innovations are more common among the small firms, while process innovations are more common among the larger ones.

More than 22 percent of the large firms in the Stockholm region have an R&D-intensity above 10 percent of their turnover. Among the small firms, having an R&D-share of at least 10 percent, more of the R&D money is used for innovation activities. The share is higher (15 percent) in the Stockholm region than in the rest of the Mälar region. One explanation to this vintage capital phenomenon is that the R&D share discloses information about the product group that the company specialises in. Raw materials and semi-finished products have a lower degree of specialisation than the products we treated in the earlier discussion. The innovation intensity, that is, the costs for innovation activity relative to turnover, is higher in the Stockholm region than in the rest of the metropolitan region.

Table 4.6 shows how the R&D share varies across industries. Not surprisingly the highest share is found in the R&D sector.

Table 4.6 R&D share and sector by region (as percentage)

R&D Share	R&D-Intensive	Knowledge Intensive	Capital Intensive	Labour Intensive	Protected	Total
Stockholm Region						
0-4 Percent	28.9	70.4	83.3	50	71.1	56.5
5-9 Percent	26.3	11.1	16.7	36.4	13.3	20.3
10 Percent or More	44.7	18.5	0	13.6	15.6	23.2
Total	(38) 100	(27) 100	(6) 100	(22) 100	(45) 100	(138) 100
Rest of Region						
0-4 Percent	30.4	56.3	66.7	83.6	83.8	66
5-9 Percent	30.4	29.7	27.8	6.5	16.2	21.3
10 Percent or More	39.1	14.1	5.6	10.9	0	12.8
Total	(23) 100	(64) 100	(18) 100	(46) 100	(37) 100	(188) 100
Metropolitan Region						
0-4 Percent	29.5	60.4	70.8	72.1	76.8	62
5-9 Percent	27.9	24.2	25	16.2	14.6	20.9
10 Percent or More	42.6	15.4	4.2	11.8	8.5	17.2
Total	(61) 100	(91) 100	(24) 100	(68) 100	(280) 100	(326) 100

The definition of innovation adopted in this part of the study is based on the concepts defined in the introductory part of the book. Product innovations are therefore defined as either substantial improvements of an existing product as regards components, quality, image or design or construction of a product completely new to the company. Process innovations are defined as substantially improved or new methods to produce goods and services through changes in the organisation of production or the machinery equipment used.

According to this definition, process innovations have occurred in 34 percent of the small and medium-sized firms (SMEs) in the Stockholm region, and in two thirds of the large firms. The corresponding figures for firms in the rest of the metropolitan region are 36 and 69 percent. In the SMEs, product innovations are more numerous, being around 50 percent in both sub-regions. Only 30 percent of the large firms stated that they had made product innovations since 1994.

Fig. 4.11 illustrates the prevalence of product and process innovations among the surveyed firms in the Stockholm region. The subdivision into sectors is again based on the sector categories most used in the Swedish industrial and innovation policy documents (see NUTEK 1997). It is in the R&D-intensive sector that we find the majority of product and process innovations. The capital and knowledge intensive sectors are also relatively active in terms of innovation compared with the labour-intensive and protected sectors. The former spend the largest share of turnover on innovation. A comparison between Swedish and foreign-owned firms show that both product and process innovations are more common among companies that are fully or partly foreign-owned.

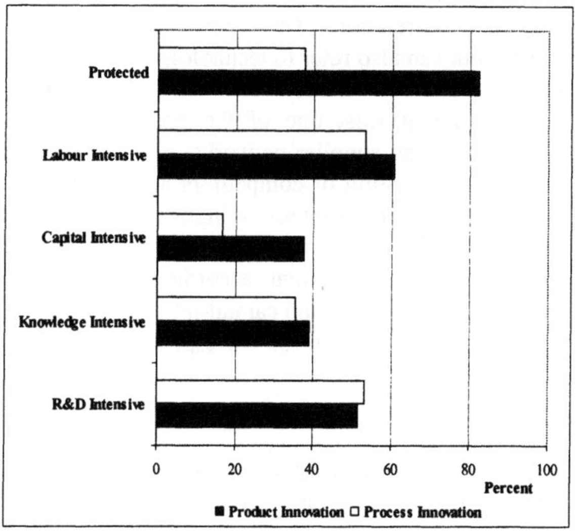

Fig. 4.11 Share of responding firms engaged in product and process innovations since 1994

The share of turnover spent on R&D focuses on the resources that the firm is devoting to innovation. The number of patent applications, on the other hand, can be seen as an indicator of the results of the innovation work. From Table 4.7 we learn that it is within the R&D intensive sector that the share of firms applying for patents is the highest.

Table 4.7 Share of firms that have applied for patents

Sector	Stockholm Region	Rest of Region	Metropolitan Region
R&D Intensive	42.1	46.2	43.8
Knowledge Intensive	31.0	37.1	35.4
Capital Intensive		45.0	37.0
Labour Intensive	20.0	13.3	15.6
Protected	12.3	20.5	15.8

C Networks and Network Formation

Although firms in all industries engage in co-operative activity, it is most prevalent in the R&D-intensive industry. Manufacturing firms producing raw materials and semi-finished products co-operate less than others. In order to be able to produce innovations, firms require the relevant information and knowledge. Information can be obtained through contacts with different actors. These may be customers, suppliers, or R&D institutions. Together, they form an open network that can be defined in this case as constituting co-operation in innovation work outside normal business relations. These networks may express themselves not only in spatial terms but can also refer to technology areas.

The level of co-operation shown in Table 4.8 is defined in terms of the number of firms that have networked through at least one of the possible business relations: customer networks, manufacturing supplier networks, producer service supplier networks, producer networks consisting of competitors and other firms, and research institutions. It is clear that the R&D-intensive industry has the most elaborate co-operation network. This holds for both the Stockholm region and the rest of the metropolitan region. It should be noted that, according to the survey results, the protected industry as well as the labour and capital-intensive sectors in the rest of the region co-operate more than firms in the corresponding sectors in the Stockholm region.

Table 4.8 Share of surveyed firms involved in co-operation in innovation work outside of normal business relations

Sector	Stockholm Region	Rest of Region	Metropolitan Region
R&D Intensive	92.9	80.0	87.5
Knowledge Intensive	81.8	78.2	79.3
Capital Intensive	60.0	95.5	84.4
Labour Intensive	48.8	72.2	63.7
Protected	56.9	66.7	61.5
Total	(129) 67.5	(196) 75.7	(325) 72.2

A subdivision of the firms by size shows that 62 percent of small firms and 87 percent of large firms in the Stockholm region are engaged in some form of co-operation. The corresponding figures for the rest of the metropolitan region are 69 and 86 percent respectively. It also emerges that firms engaged in the production of raw materials and semi-finished products engaged in less co-operative activity that those that produce more complex products. This holds primarily in the Stockholm region. The classical manufacturing firms have also extensive co-operation if they are located in the rest of the metropolitan region.

Fig. 4.12 shows the relative importance of different technology fields for innovation in manufacturing companies. Firms were asked to state the importance of these fields on a five-point scale. As can be seen, process technology, materials technology, environmental engineering, and information and communications technology were given the highest ratings.

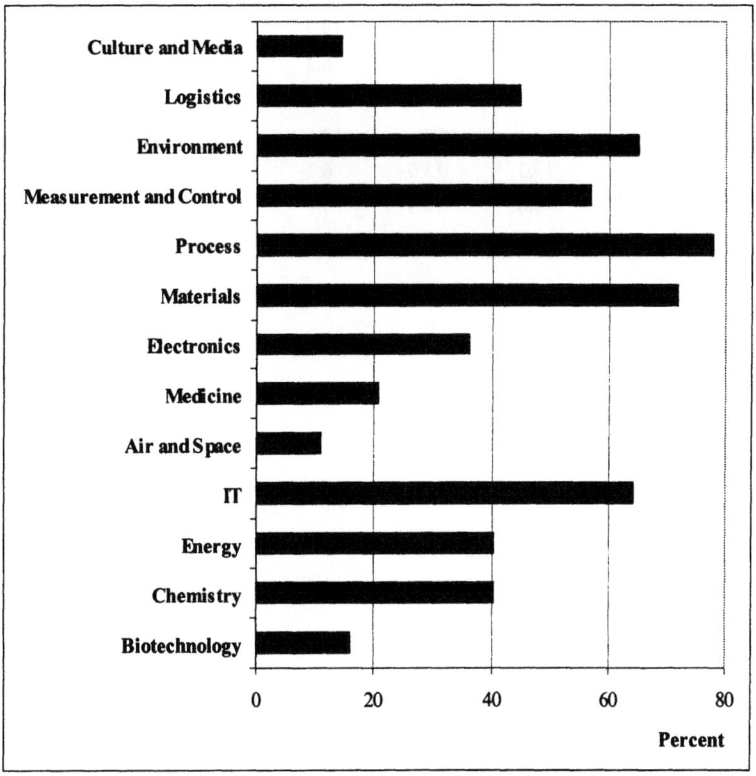

Fig. 4.12 Relative importance of technology fields for innovation

Table 4.9 gives an idea of the share of firms in the five industrial sectors that engage in co-operation in the various technology areas. Some of the latter have been pooled in order to increase the number of supporting observations. The most distinctive is the capital-intensive industry, where firms in the Stockholm region are particularly inclined to seek their technologies in the biotechnology, chemistry and energy fields, as well as in the environment and logistics. As a rule, co-operation within smaller technology areas is regarded as less important. The differences within the metropolitan region are rather small. It should be noted that firms belonging to the non-protected sectors do not see culture and media technology as an important field of co-operation.

Table 4.9 Share of firms by sector that have co-operated in different technology areas in the Stockholm region

Technology Field/Industrial or Manufacturing Sector	R&D Intensive	Knowledge Intensive	Capital Intensive	Labour Intensive	Protected
Biotechnology and Energy	13.1	11.5	30.2	19.4	12.9
Information, Air and Space	18.2	10.1	19.5	7.5	17.3
Medicine and Electronics	20.4	5.3	6.3	7.5	5.4
Materials	13.1	20.7	nr	19.4	10.8
Production and Process Control	13.1	23.1	12.6	19.4	16.2
	15.3	13.0	6.3	7.5	4.9
Environment and Logistics	6.6	16.3	25.1	19.4	21.6
Culture and Media	-	-	-	-	10.9

Note: nr = no response

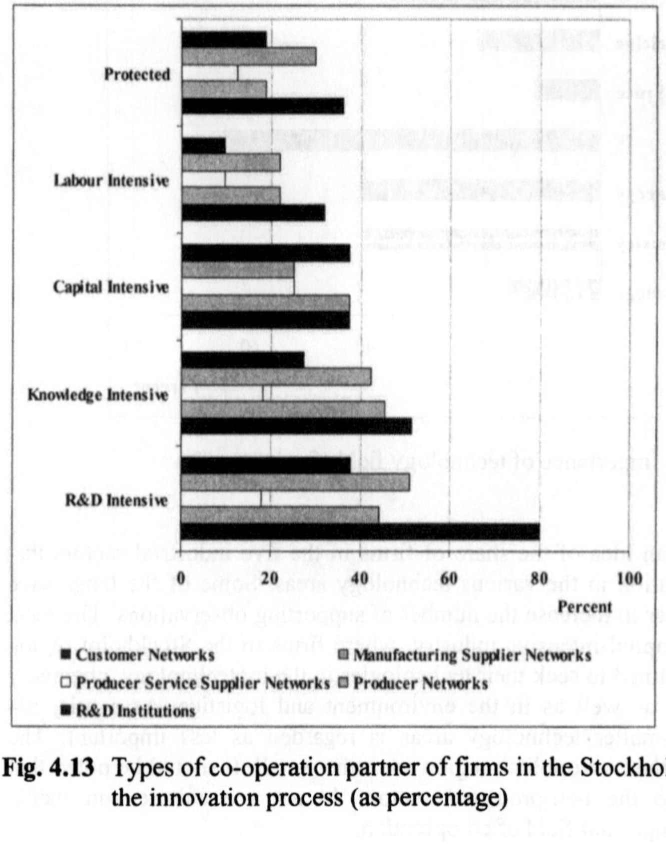

Fig. 4.13 Types of co-operation partner of firms in the Stockholm region during the innovation process (as percentage)

In Fig. 4.13 the co-operation partners have been subdivided into five classes: customers, suppliers, business service firms, other firms such as competitors, and R&D institutions. This gives an indication how firms in the Stockholm region distribute their contacts during the innovation process. It emerges clearly that manufacturing firms co-operate primarily with their customers, who are the partners in the strongest position to put claims on the products to be delivered. In the Stockholm region it is also common to co-operate with business services firms, except in the capital-intensive branch where firms also tend to promote co-operation with suppliers. The pattern is more dispersed in the rest of the metropolitan region. The most common sources of knowledge for product innovations are in fact usually customers, competitors, and fairs and exhibitions. For process innovations it is suppliers, fairs and exhibitions, and professional literature that dominates the networking activities.

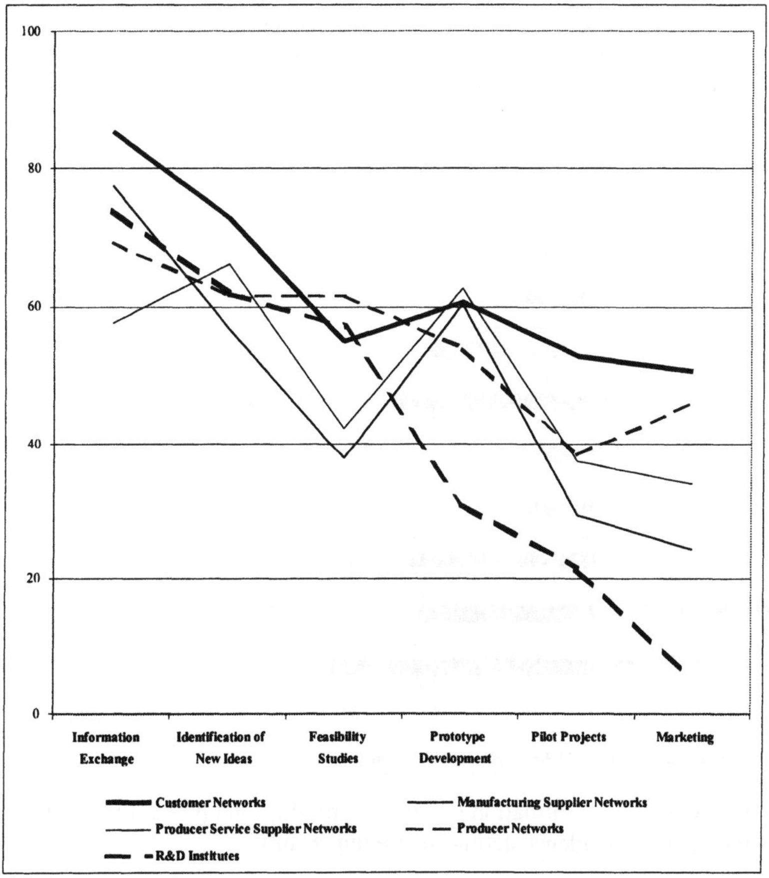

Fig. 4.14 Co-operative activity of firms in Stockholm region by type of partner during different stages of the innovation process (as percentage)

To reveal what kind of co-operation partners firms have in different industries, a more thorough analysis of the data is needed. It is also important to know whether certain types of partners are more common in some stages of the innovation process than others and how firms are able to gather knowledge for their innovation work. It is clear from Fig. 4.14, that the relative importance of the fields of co-operation changes as the innovation process proceeds. The figure shows the share of firms having co-operated with different firms during the stages of the innovation cycle.

During the general information exchange phase, Stockholm-based firms are more oriented towards R&D institutions than firms in the rest of the metropolitan region. The latter co-operate primarily with their customer networks. On the other hand, during the feasibility study phase, the firms in the rest of the region work more with manufacturing suppliers and producer service providers than firms in the Stockholm region.

Fig. 4.15 illustrates the most important sources of information for product and process information, respectively. It reveals that the most important sources of information for product innovations are customers, fairs and exhibitions, competitors, and suppliers in that order. The relative ordering is almost the same in both parts of the metropolitan region.

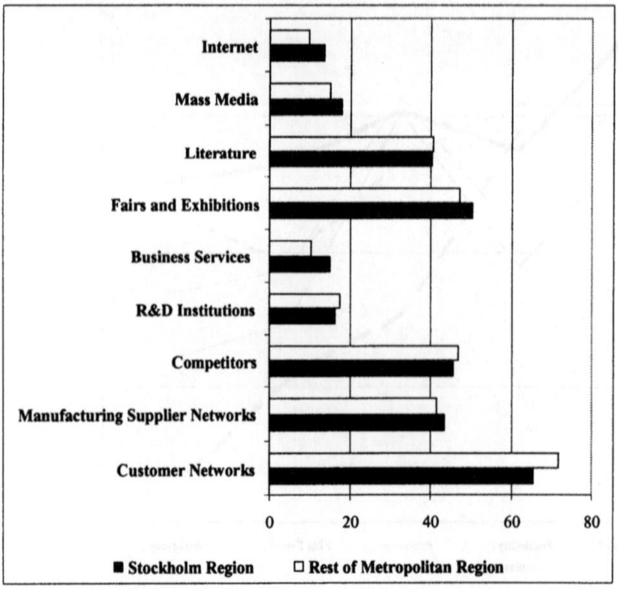

Fig. 4.15 Importance of information sources for product and process innovations (share of respondents stating source important)

For process innovations, the most common source of information is not customers but suppliers. After that come fairs and exhibitions, and professional literature. Evidently, because of requirements for the products that they buy, firms are more involved in the production processes of their suppliers than in the way production is organised among their customers. It should be noted, of course, that the firm in question will be acting as the customer of its supplying firms.

Fairs and exhibitions are important for both product and process innovations. This shows that firms need meeting places outside the interaction with customers and suppliers to exchange their views and obtain information about new products, services and technologies.

D The Location of Innovation Co-operation Partners

The geographical distribution of the innovation networks can be measured in several different ways. One is to simply describe the location of the different partners stated by the firms. The survey allows us to characterise the level of intensity of the co-operation for the different partners.

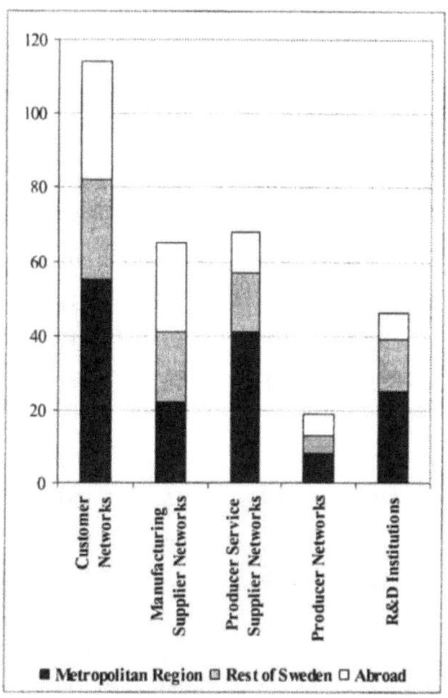

Fig. 4.16 Number of very intensive co-operation partner contacts by type and geographical area for manufacturing firms in the metropolitan region

As we can see from Fig. 4.16, the co-operation partners of firms in the Stockholm region are primarily found in the region itself or in the rest of Sweden. On the other hand, the geographical spread of the pattern of co-operation for manufacturing firms in the rest of the metropolitan region has a different appearance; co-operation partners are spread quite evenly across Sweden as a whole.

We now develop a measure of co-operation that also takes into consideration the intensity of the activity. The approach we have used is to weight the number of contacts with the intensity of those contacts, as stated by the respondents, on a three-degree scale. This method allows us to give a comparable treatment to a variety of different contact intensity patterns as stated by the responding firms. In the analysis presented in Table 4.10 the intensity of co-operation is described by counting only those contacts which the firm has declared to be very intensive in the weighting procedure.

It can be seen clearly from the table that the contacts in the home region are the most important for firms located in the Stockholm region. The firms in the rest of the metropolitan region have their most intensive contacts with actors in Swedish regions outside the metropolitan area. As the table makes evident, the difference between the two weights is be important in some cases, but much less so in others. A case where there is a significant difference is the manufacturing supplier networks in the Stockholm region.

Table 4.10 Importance of co-operation in different geographical locations measured as share of firms stating link to be very important

Network Type	Firm Location/ Partner Region	Stockholm Region	Rest of Region	Rest of Sweden	Abroad	All Regions
Customer Networks	Stockholm	73	4	8	15	100
	Rest of Metro	10	15	63	12	100
Manufacturing	Stockholm	29	2	30	39	100
Supplier Networks	Rest of Metro	3	9	41	47	100
Producer Service	Stockholm	79	0	11	10	100
Supplier Networks	Rest of Metro	10	45	39	6	100
Producer Networks	Stockholm	11	0	33	56	100
	Rest of Metro	39	15	23	23	100
R&D Institutes	Stockholm	67	0	22	11	100
	Rest of Metro	45	20	30	5	100

A summary of some of the distinguishing features of the co-operation networks between manufacturing firms, their customer networks and R&D institutions is provided in Table 4.11.

Table 4.11 Patterns of co-operation of manufacturing firms with customer networks and R&D institutes (share of firms stating co-operation to be intensive or very intensive)

Activity	Customer Networks	R&D Institutions
Innovation Co-operation Exists	*56*	*25*
Pre-Competitive Stage Co-operation		
Information Exchange	21	27
Identification of New Ideas	25	24
Research and Development	18	25
Competitive Stage Co-operation		
Prototype Development	30	14
Pilot Projects	18	12
Market Introduction	10	7
Spatial Proximity Important		
In the Pre-Competitive Stage		
Information Exchange	80	68
Identification of New Ideas	68	51
Research and Development	53	50
In the Competitive Stage		
Prototype Development	61	27
Pilot Projects	45	20
Market Introduction	46	6
Co-operation Never Important	*33*	*38*
Most Intensive Co-operation Region		
Stockholm Region	34	52
Rest of Metropolitan Region	10	14
Rest of Sweden	42	27
Abroad	14	7
All Regions	100	100

The table shows that co-operation is more than twice as frequent for customer networks than for R&D institutes. In both relationships, however, the co-operative activity tends to be more important in the early stages of the innovation process. The decline along the innovation cycle is more pronounced for contacts with R&D institutions. Spatial proximity is more important with customers, when stated to be of importance. Here too, the need to have partners close in space is more pronounced in the first stages of the innovation process. Finally, partners in the Stockholm region are more important in the R&D contacts, but customer networks seem to be spread across the whole country.

4.5 The Producer Services Sector

A Structure of the Sector

The Stockhom metropolitan region has a specialisation profile which is particular within the Swedish economy, and even at the European scale. As noted earlier, the share of employment in private and public services was higher at the beginning of the 1990s than in any other functional urban region in Europe (see Snickars 2000). Fig. 4.17 gives a summary of the specialisation profile in Sweden as a background to the presentation of the results as regards the innovation potential and innovation networks of the producer service sector in the region. The specialisation is given in terms of the share of the employment in the total metropolitan region, and the Stockholm region as the metropoltian core, as compared to the share in the country as a whole.

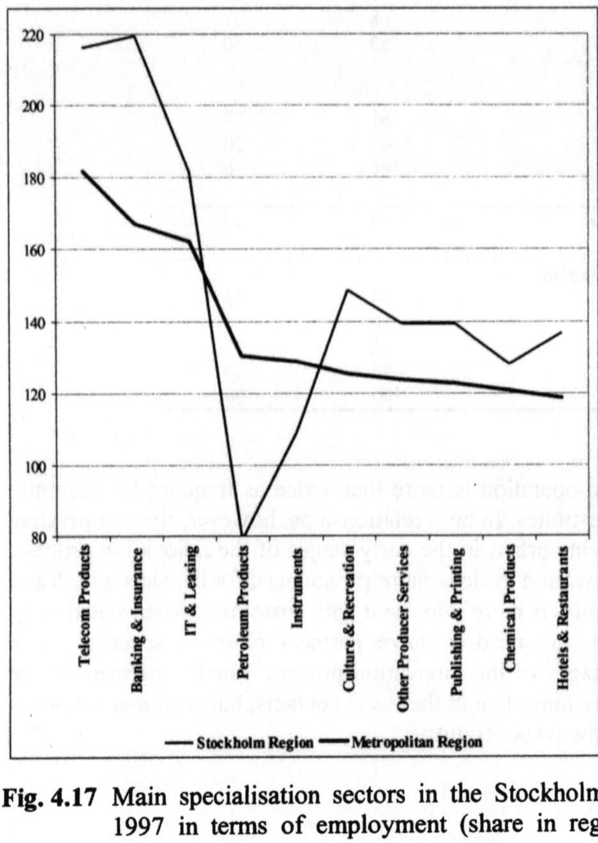

Fig. 4.17 Main specialisation sectors in the Stockholm metropolitan region in 1997 in terms of employment (share in region relative to Sweden, average for Sweden as a whole = 100)

It is clear from the figure that the metropolitan region is indeed a service-producing region. Of course, it should be pointed out that the region houses the international telecom corporation Ericsson, as well as petroleum refining capacity at national scale. It also has a substantial specialisation in the instruments industry. Besides these manufacturing profiles, the region plays a banking and insurance function, is an IT and leasing region, and has a large share of the jobs in the national culture and media industries as well as producer services in general. In brief, the growth in jobs and economic capacity in the metropolitan region comes basically from the sectors listed as the main specialisation in Fig. 4.18.

The innovation study involved all firms in manufacturing and business services with at least 10 employees in 1996. Furthermore, questions have been asked about innovation behaviour of all R&D institutions in the Stockholm metropolitan region. The sample of manufacturing firms includes some 450 manufacturing firms (27 percent), about 350 establishments providing business services (26 percent) and 175 R&D institutes (52 percent). The response rate allows statistical representativeness at a macro level, but does not permit detailed analyses for subgroups of sectors and subregions.

Table 4.12 contains a summary of the producer service sectors included in the questionnaire sample in the Stockholm region. The list includes the culture and media industry, which was covered in the Stockholm version of the innovation survey. Culture and media firms comprise 12 percent of firms and eight percent of the jobs, inplying that they are somewhat smaller than the average producer service firm. In the Stockholm study, the finance and banking sectors have also been included among producer services. These sectors comprise four percent of the firms and as much as nine percent of the jobs in the producer service sector as a whole. All firms and institutions were asked to assess their contacts with the culture and media, and finance and banking, industries. Furthermore, the respondents were asked to state the role of capital market factors, as well as cultural activities and cultural environments as location factors and carriers of locational value for innovation activities. The table compares the Stockholm region and the rest of the metropolitan (Mälar) region.

Table 4.12 Regional composition of producer services sector and share of firms located in Stockholm region

Sector	Stockholm Region	Rest of Metro Region	Stockholm Region Share
Finance	4	7	68
Technical Consultancy	8	18	63
Computer Software	29	26	81
Business Consultancy	34	31	81
Market Research and Advertising	13	6	90
Culture and Media	12	12	75
All Producer Services	100	100	79

The survey contains responses from 100 IT firms and some 35 technical consultancy firms. More than 120 firms belong to the business consultancy sector. There were some 20 financial institutions and around 40 firms in the culture and media industry. All sectors have both a production systems orientation and an orientation towards final consumption. The actual number of jobs in the producer services sector in the sample was relatively small. We see from the table that the finance, and culture and media, sectors have rather lower concentrations in the Stockholm region than the average.

A summary of selected characteristics of the producer services firms surveyed in the Stockholm region is provided in Table 4.13. In the sample that forms the basis for the survey there were rather more single establishment firms than multiple establishments. More than half of the firms employed not more than 20 persons. The innovating firms are more concentrated among the multiple establishment ones than the average, and more in large firms than small firms. The sector consists of technically-oriented producer firms as well as business-oriented ones. Examples would be IT firms and marketing firms, respectively. The business-oriented enterprises are more often single establishment firms, whereas the technically-oriented ones more often belong to a corporate structure with more than one establishment. It comes as no surprise to find that the business-oriented firms are also on average smaller than the technically-oriented ones.

Table 4.13 Selected characteristics of sample of producer service firms in the Stockholm region (number of firms and percentage of total)

	All Firms		**Innovating Firms**[a]		**Non Innovating Firms**[b]		**Technically-Oriented Firms**[c]		**Business-Oriented Firms**[d]	
Corporate Status										
Single Plant	122	57	86	53	36	69	61	50	61	66
Multi Plant	92	43	76	47	16	31	60	50	32	34
Main Plant	*57*	*27*	*50*	*31*	*7*	*14*	*36*	*30*	*21*	*22*
Branch Plant[e]	*35*	*16*	*26*	*16*	*9*	*17*	*24*	*20*	*11*	*12*
Total	214	100	162	100	52	100	121	100	93	100
Employment Size										
≤ 19	128	55	87	49	41	71	64	48	64	62
20 – 49	62	26	50	28	12	20	33	25	29	28
50 – 249	34	14	30	17	4	7	25	19	9	9
≥ 250	11	5	10	6	1	2	10	8	1	1
Total	235	100	177	100	58	100	132	100	103	100

Notes: **a** Producer service firms introducing new services and/or organisational innovations
b Producer service firms not introducing new services or organisational innovations
c Producer service firms in computer software and technical consultancy
d Producer service firms in business consultancy and market research/advertising
e Branch establishment or subsidiary

Very few producer firms spend nothing on R&D, while one quarter spend more than 3.5 percent (see Table 4.14). This breakpoint is somewhat lower than the level of spending at the national level as a share of GNP. The largest firms in the sample do not, in fact, spend a larger share of their turnover on R&D activities than the small ones. According to the table, almost one third of the smallest firms spend on average more than 3.5 percent of their turnover on R&D activities.

Table 4.14 R&D expenditure of producer service firms by size classes (number of firms and percentage of average turnover over last three years)

	Firm Size by Number of Employees									
	≤ 19		20 – 49		50 – 249		≥ 250		Total	
	no.	%[a]	no.	%[a]	no.	%[a]	no.	%[a]	no.	%[a]
No R&D	3	4	0	0	0	0	0	0	3	2
b	*100*		*0*		*0*		*0*		*100*	
0.1-3.4 %	49	64	35	82	20	80	8	89	112	73
b	*44*		*31*		*18*		*7*		*100*	
3.5.7.9 %	13	17	7	16	4	16	1	11	25	16
b	*52*		*28*		*16*		*4*		*100*	
≥8 %	11	15	1	2	1	4	0	0	13	9
b	*85*		*8*		*7*		*0*		*100*	
Total	76	100	43	100	25	100	9	100	153	100
b	*50*		*28*		*16*		*6*		*100*	

Notes: a Percentage of column total
b Percentage of row total

B Innovation Activities of Producer Service Firms

It is sometimes asserted that innovativeness is concentrated in newly established firms, which thrive on the upward curve of the product cycle. This observation is supported by the empirical evidence in Table 4.15. While the share of innovative firms, according to our definition, is some 15 percent among the firms existing since 1970, among those set up in the 1990s it is 40 percent.

Table 4.15 Product innovation intensity of firms by vintage class (firms with product innovations since 1994)

	Year Establishment Founded									
	Before 1970		1970-79		1980-89		Since 1990		All	
	no.	%[a]	no.	%[a]	no.	%[a]	no.	%[a]	no.	%[a]
Non-Innovative Firms	11	18	3	5	20	33	27	44	61	100
Innovative Firms	25	14	19	11	61	35	70	40	175	100
Total	36	15	22	9	81	34	97	41	236	100

Note: a Percentage of row total

Table 4.16 Producer service firms with product innovations (1994-1996)

Introduction of Product Innovation	Single Establishment		Main Plant		Branch Plant		Total	
	no.	%[a]	no.	%[a]	no.	%[a]	no.	%[a]
No Innovations	36	69	7	14	9	17	52	100
New Services	51	51	36	36	14	14	101	100
Improved Services	52	50	38	36	15	14	105	100
Organisational Innovations	46	54	25	29	14	17	85	100

Note: a Percentage of row total

Table 4.16 adds to these observations by showing how the innovation activities are distributed among firms of different corporate structure. It is evident that, all things equal, product innovations are more common among the establishments which belong to a larger corporate structure. Some 70 percent of the single establishment firms made no product innovations during the period 1994-96. Organisational innovations seem to be rather more prevalent among single establishment firms than service innovations.

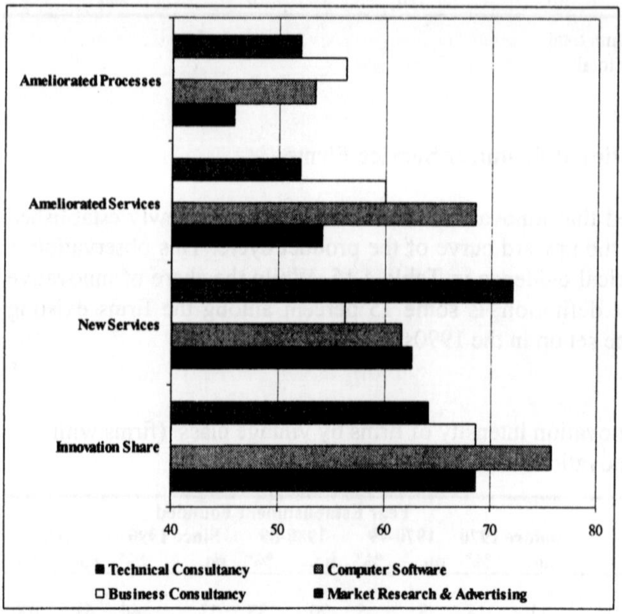

Fig. 4.18 Share of firms participating in various forms of innovation activity (percentage of responding firms)

The above considerations concerned the role of producer service firms in innovation activities in manufacturing. The firms themselves are also involved in making their own production processes more effective and improving the quality of their services. The survey attempted to define an innovation concept applicable to the service industries, by defining three categories: new services, substantially improved services, and substantially improved service production processes. As a starting point, firms were asked whether they felt they were involved in innovation activities in the home base. Fig. 4.18 presents a summary of the results for four sub-sectors of the producer services branch. The first observation is that around two out of three firms stated that they are engaged in innovation activity in their own production. The share is three out of four for the computer software industry, which indicates the dynamic nature of that industry.

Three out of five firms have introduced new service products in the last two years, with a higher share in the market research and advertising sector. The innovation rate is somewhat lower, but still around 50 percent as regards improved services and production processes. Again, the computer software industry seems rather more dynamic than the others. A general observation is that the producer services firms see innovation in their own firm mostly arising in the process of service product development and to a somewhat lesser degree in the search for a more efficient production process. Innovation is thus more market expansion driven than cost effectiveness induced, a phenomenon typical of firms on the upward curve of the product life cycle.

C Networks and Network Formation

Producer service firms engage in innovation activities in their own right in order to achieve higher efficiency in their production. At the same time, they play the role of service providers to the manufacturing sector, both as regards outsourced support activities for current production and as participants, quite often acting as the driving force in innovation activities in manufacturing. Our analysis of the role of the producer service sector in the innovation system of the Stockholm region and its hinterland includes networks which involve manufacturing firms. These networks can be of many types, among which innovation support is only one function. As we can see in Table 4.17, one producer service firm in two stated that it was engaged in process and organisational innovations in the manufacturing sector. Somewhat fewer, around 40 percent, are involved in product innovation co-operation and support the extension into new markets for the manufacturing sector.

Focusing on the role of product and process innovation, we find that firms' technical consultancies are the service firms most engaged in product innovation (two thirds state that this is the case). Computer software companies participate to a large degree in the process innovation activities with the manufacturing sector. At the same time, there seems to be a strong correlation between the business consultancy sector and organisational innovation, and between the market research and advertising sector and new market extensions.

Table 4.17 Types of innovation support for manufacturing firms by producer service firms (percentage)

Innovation Type	Technical Consultancy	Computer Software	Business Consultancy	Market Research & Advertising	All Producer Services
Product	66	42	40	29	42
Process	46	60	56	22	50
Organisation	40	51	62	31	48
Market Extension	31	37	38	43	38

Collaboration with industrial clients is facilitated by some factors and hindered by others. In general, frequent personal contacts were considered the most important factor in collaboration with industry. Business consultancy respondents rated personal contacts particularly highly. Other factors, indicated in Table 4.18, were regarded as almost as important, the lowest rating being given to spatial proximity (mentioned by less than one out of five respondents). On the whole, however, the differences are not significant among the sectors within the producer services industry.

Table 4.18 Factors for successful collaboration with industrial clients (firms stating factor to be important or very important)

Collaboration Factors		Technical Consultancy		Computer Software		Business Consultancy		Market Research & Advertising		All Producer Services	
		no.	%[a]	no.	%[a]	no.	%[a]	no.	%[a]	no.	%[a]
Spatial Proximity		20	21	47	18	19	17	14	18	100	18
	b		*20*		*47*		*19*		*14*		*100*
Frequent Personal Contacts		28	30	73	28	37	33	23	30	161	30
	b		*17*		*45*		*23*		*15*		*100*
Similar Qualifications		22	24	66	26	31	27	19	25	138	25
	b		*16*		*48*		*23*		*13*		*100*
Knowledge of Clients		25	25	72	28	27	23	20	27	144	27
	b		*17*		*50*		*19*		*14*		*100*
All Factors		95	100	258	100	114	100	76	100	543	100
	b		*18*		*48*		*21*		*13*		*100*

Notes: a percentage of column total
b percentage of row total

Producer service firms give innovation support to manufacturing firms through a variety of forms of contact. The factors indicated in the above table facilitate contacts and make them easy to maintain, but are not necessarily used to actually establish the contact. This information is given in Table 4.19 for the same four sub-sectors of the producer service industry. The table shows that by far the most important contact form is through former colleagues, or through literature and references. These two forms were stated as being of prime importance in almost 60 percent of the responses. Contact agencies, on the other hand, were rated to be unimportant for the firms in establishing their innovation partnership contacts.

Looking at the table from another viewpoint, it is clear that the computer software companies have many more forms of contact than other producer services firms. To some extent this may be due to the fact that more software companies responded to the question. There were around 120 business consultancy firms in the group of responding firms, some hundred computer-software firms, and some 40 technical consultancy firms and market research and advertising companies each (we exclude here the finance, and culture and media firms). Compensating for this we see that the market research and advertising firms as well as the computer software ones state more contact channels than the others as important or very important. The business consultancy firms seem to be less dependent on active networking than the other firms.

Table 4.19 Establishments of contacts through various channels (firms stating factor to be important or very important)

Contact Channels	Technical Consultancy		Computer Software		Business Consultancy		Market Research & Advertising		All Producer Services	
	no.	%[a]	no.	%[a]	no.	%[a]	no.	%[a]	no.	%[a]
Advertising	10	13	30	12	19	15	30	26	89	15
Conferences	8	10	25	10	14	11	6	5	53	9
Fairs	3	4	13	5	2	2	0	0	18	3
References	22	28	73	28	32	25	31	27	158	27
Contact Agencies	2	3	6	2	5	4	5	4	18	3
Former Colleagues	28	35	76	29	43	33	32	28	179	31
Former Employees	6	7	35	14	13	10	12	10	66	11
Total	79	100	258	100	128	100	116	100	581	100

Note: a Percentage in percent of responses

Table 4.19 also shows that the relative importance of the contact channels is about the same for all sub-sectors, except market research and advertising. It is natural that such an industry uses far more advertising in its contact establishment work than the others. For firms in this sub-sector of the producer services industry, former colleagues and references are even more important. The result indicates that the publication of written material is still a very important means of establishing contacts in the field of innovative activity.

There are also typical obstacles to co-operation between producer services firms and their innovation partners. These barriers influence the rate of economic growth in the region. They relate both to deficiencies in the supply of production factors, and problems typical of innovation activities as a whole, especially when these involve co-operation between independent companies, and include risks, implementation difficulties, and the fear of partners not fulfilling commitments.

Fig. 4.19 shows the replies of respondents to the question about hindrances to innovation in the four subsectors of the producer service industry in the Stockholm metropolitan region. The main hindrance seems to be the lack of skilled labour, which is seen as a barrier to further innovation activity by almost three firms in five. It is more significant than capital supply limitations. On the other hand, problems associated with co-operation with firms and R&D institutions were not in general seen as important barriers. They were considered less important than the difficulties caused by the fact that innovation activity is a risky business in itself with substantial costs and uncertain payoffs. The picture provided suggests that internal factors are seen as more serious hindrances than external ones relating to the regional production environment.

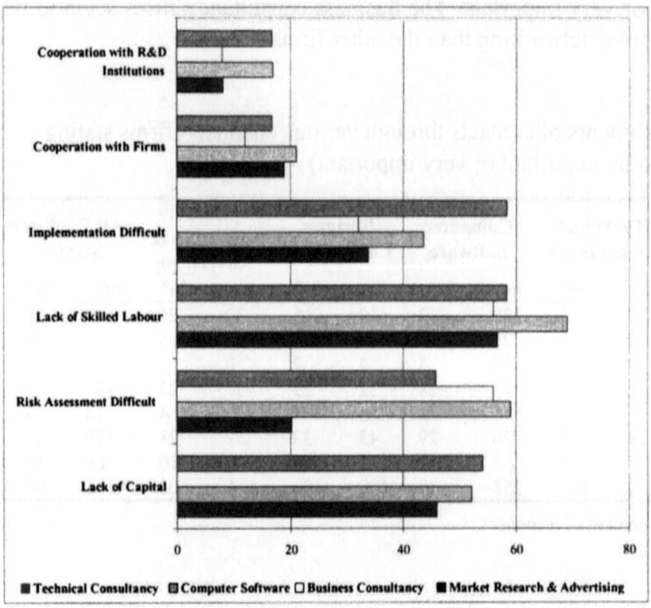

Fig. 4.19 Hindrances to innovation activity in producer services sector in the Stockholm metropolitan region (share of respondents stating hindrances to be large or very large)

The differences between the four sub-sectors in their statements relating to hindrances are relatively insignificant. This might partly be due to the fact that the replies regarded both hindrances in the support given by producer service firms to the innovation activities in manufacturing firms, and hindrances for due to their own internally induced innovations. Computer software firms find the supply of skilled labour as a bigger hindrance than the other firms. Technical consultants are often specialists in risk management and consequently see this as less of a problem than the other firms. The market research and advertising companies find that implementation problems slow down their innovation activities more than the other firms. For business consultants, the lack of capital is somewhat less pressing than for the other firms. The impression is that the hindrance profile reflects the inherent nature of the production system in which the four firm types work more than being indicators of problems in the regional production milieu.

D Location of Innovation Co-operation Partners

Producer service firms in the Stockholm metropolitan region co-operate with manufacturing partners, partners in the producer service industry itself, and also with R&D institutions. They were asked a number of related questions in the survey about the pattern of co-operation, the number of partners, the intensity of the co-operation, the geographical location of the partners, and the sector of the economy with which the interaction takes place. Table 4.20 illustrates the result regarding the last type of co-operation. The table reveals that the average number of partners is around 30, the bulk of these being in the manufacturing sector. The other sectors of the economy are insignificant when it comes to partnerships in innovation activity in the metropolitan region. It also emerges that the technical consultants have particularly large networks, with as many as 80 partners mentioned. It seems likely that in this case the firms responding have been rather wide in their interpretation of innovation. By contrast, the culture and media sector has rather a different co-operation pattern with far fewer firms in their partnership networks on average.

Table 4.20 The average number of innovation partners in different sub-sectors for various components of the producer services sector (percent)

	Manu-facturing	Producer Services	R&D Institutions	All Partners
Business Consultancy	21	1	0.5	22.5
Computer Software	25.5	2.0	0	27.5
Market Research & Advertising	22	4	0	26
Technical Consultancy	82	5	1	88
Culture & Media	12	0.5	0	12.5
Finance	20.5	3	5	28.5
All Producer Services	27.5	2	0.5	30

The picture given above is supplemented by Table 4.20, which shows the share of firms in the co-operation profile of the various sub-sectors by region. On average, about half of the co-operation partners are located within the metropolitan region. The share is particularly large for the business consultancy sector, where four out of five links lie within the region. The technical consultancy sector has a very different profile, as only one third of the partners are found within the metropolitan region. The market research and advertising sector resembles business consultancy firms, whereas the regional distribution of the pattern of co-operation in innovation work in the computer software sector has more similarities with technical consultancies. One can say in general that international partnerships are relatively rare, with less that one partner out of five to be found abroad.

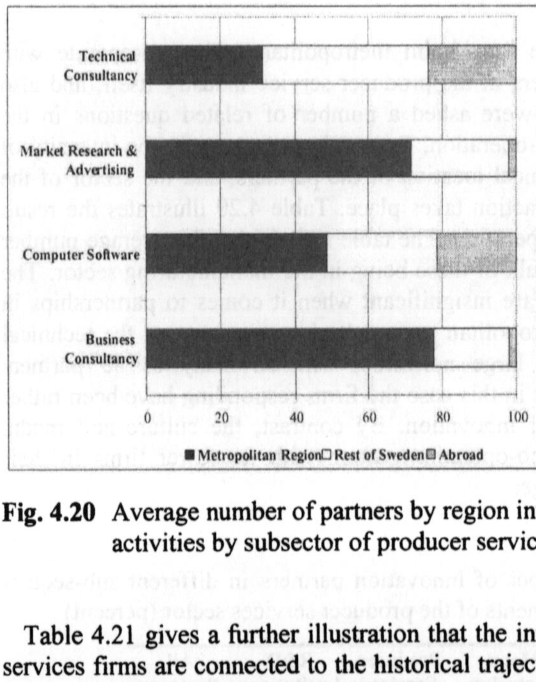

Fig. 4.20 Average number of partners by region involved in innovation activities by subsector of producer services

Table 4.21 gives a further illustration that the innovation networks of producer services firms are connected to the historical trajectory of the employees. We saw above that contacts with earlier colleagues were considered the most important for innovation networks. The following table shows that two respondents out of five state that earlier positions at research institutes have not been central to their current innovation networks. Contacts generated during doctoral education do not seem to count very much either. Instead, it is earlier business positions that are seen to generate the momentum for current innovation networking activites.

Among the contacts mentioned as important, business contacts dominate. In intraregional co-operation, business contacts are stated by more than one in two respondents as being important or very important sources of co-operation for current innovation. The figure is even higher for the international contact networks. This indicates a certain deviation between the education and R&D system in the metropolitan region, on one hand, and the industrial innovation system on the other. It seems that people who are actively involved in networking for innovation work do not use their contacts with earlier higher education opportunities as frequently as one might expect on a priori grounds. The result would seem to indicate a mismatch between the industrial innovation system and the R&D system. This might in part be a reflection of the simple fact that the respondents were mainly involved in the managerial activities of producer service firms, and probably their networks are not as strongly related to the university system as one might believe. It also raises the question to what extent networking can bridge this gap.

Table 4.21 Relevance of personal contacts from earlier career activities of employees (rated as important or very important)

	Metropolitan Region		Rest of Sweden		Abroad		Actor not Important	
	no.	%[a]	no.	%[a]	no.	%[a]	no.	%[a]
University Studies	39	27	49	31	9	22	70	21
Doctoral & Post Doctoral Job	23	16	31	20	5	12	101	31
Research Institutes	9	6	19	12	3	7	129	39
Business Position	74	51	59	37	25	59	28	9
Total	145	100	158	100	42	100	328	100

Note: a percentage of column total

4.6 The Science & Research Sector

A General Features of the Sector in the Region

During the 1980s, co-operation between manufacturing industry and university became the focus of increasing attention. There were several reasons for this. One was the discovery of a growing group of industries referred to as high technology. These industries had one common denominator – a growing dependence on knowledge, and especially research and development. Universities became therefore important location factors (Andersson 1985; Anderstig and Hårsman 1986; Rogers and Larsen 1984; Hall 1987). Regions with a strong university tradition were regarded as having a much better opportunity of attracting high technology industry than regions without. A plethora of empirical evidence was produced to support this notion. The development of high technology districts such as Silicon Valley and Route 128 in Boston in the early eighties indicated that the presence of well-established universities was one of several conditions for growth of high technology industry. But links between industry and university appeared to be difficult to detect, and proximity to fundamental research in universities failed to prove significant (Hall and Markusen 1985, Markusen, Hall and Glasmeier 1986; Malecki 1991). Other studies also indicated additional important factors for high technology expansion both in California and Boston, where growth to large extent was stimulated by the federal defence budget (Kuhn 1982).

More recent studies show strong evidence of a spatial relationship between the distribution of university capacity and high technology (Varga 1998). But the university effect is multidimensional and must be evaluated in at least two major ways (Florax 1992). Firstly, there is an indirect impact which can be compared to certain types of government expenditure in bringing positive employment effects to a region. The second type of effect has a direct consequence for high technology firms. In this case, a university either attracts companies into a region or is a condition for high technology firms' spin-off. The latter involves what is often called technology transfer. It consists of several sorts of co-operation, both informal and formal, which include information exchange, seminars, scientific publishing, or industrial associates programmes.

The Stockholm metropolitan region innovation survey indicates that customer relations and customer networks are more important than industry-university co-operation for the metropolitan innovation system. The most crucial regional innovation network exists in connection within business transactions. These results imply that market pull is more important than technology push. In this section, we shift the perspective from business to university and R&D institutions. The data from universities and research institutes was collected from some 170 different production units. This figure includes both universities and private research institutes.

Table 4.22 provides an overview of the results, according to the scientific field and institutional segment concerned. The coverage of the survey is substantial. We are therefore justified in considering the results representative of the behaviour and views of R&D institutions in the Stockholm metropolitan region.

Table 4.22 Response patterns and response rate of responding research units

Characteristic of Firm	Existing Firms 1997	Respondents 1997	Response Rate (in %)
Scientific Field			
Biotechnology, Chemistry, Medicine	84	77	92
Mathematics, Informatics, Physics	39	21	54
Electrical and Mechanical Engineering	45	22	49
Economics	12	12	100
Social Sciences and Geosciences	47	38	81
All	226	170	75
Legal Status			
University Departments	60	55	92
Other R&D Institutions	146	115	79
All	226	170	75
Geographical Location			
Stockholm Region	139	83	60
Rest of Metropolitan Region	87	87	100
All	226	170	75

B Innovation Activities of R&D Institutes

Information on co-operation among R&D institutes and their co-operation with business firms is shown in Table 4.23. There is no clear indication which type of network is most important. Networking characteristics clearly vary between different science fields. For example, whereas the most important form of co-operation for biotechnology, chemistry and medicine is with universities, for electrical and mechanical engineering it is with firms. This pattern indicates that theoretically-oriented science fields have less direct importance for business. It is evident for such institutions that co-operation between research institutes is more important than co-operation with business firms.

Table 4.23 Selected external co-operation characteristics of research institutes in the Stockholm metropolitan region in 1997

	Among R&D Institutes			Between R&D Institutes and Firms		
	A	B	C	A	B	C
Scientific Field						
Architecture, Construction, Surveying	8	89	7	8	89	88
Biotechnology, Chemistry, Medicine	49	71	35	40	58	33
Mathematics, Informatics, Physics	13	68	10	14	74	56
Electrical and Mechanical Engineering	14	67	11	19	90	58
Economics, Social and Geosciences	20	51	9	8	21	60
Location of Co-operation Partners						
Metropolitan Region	103	66	86	79	50	76
Rest of Sweden	69	44	39	72	46	64
European Union	91	58	69	82	52	71
Central and Eastern Europe	87	55	54	60	38	62
Rest of the World	73	46	45	58	37	48
All Research Institutes	112	71	79	97	62	49

Notes: **A** denotes number of co-operating research institutes
B denotes percentage of co-operating research institutes out of total of respondents
C denotes percentage of research institutes reporting intensive or very intensive co-operation, i.e. indicating at least three regions with intensive contacts or two regions with very intensive contacts.

The data presented in the table is not weighted with the number of scientific staff or research budget, which means that each element of co-operation is counted as one link regardless of whether the exchange contains a small or a very large amount of money or research personnel. Except for the exchange of general information, co-operation is most intensive during early phases of the innovation process. Measured in this way, it emerges that a majority of R&D institutes have co-operation with industry (as many as 62 percent co-operate with private firms, see Table 4.23)

Activities classified as generating ideas and preliminary studies represent more than half of the exchanges. At the end of the innovation process, i.e. in the phase of market introduction, co-operation reaches its lowest level. It can be seen that in the early phases of the innovation process, geographical proximity between university and industry is important, while at the time for introduction to the market, location has hardly any importance. It is also noteworthy that in 24 percent of the cases, geographical proximity is stated to have no significance. Thus, universities in the Stockholm metropolitan region co-operate with firms at local, regional, national and international level. Some 30 percent of the co-operation with industry takes place outside of Sweden. Whereas the national level co-operation accounts for 25 percent, at the local and regional level it is 46 percent. This means that national and international links are slightly more important than local and regional networks.

Even if universities are far from being the only condition for high-tech economic growth, it is obvious that the industry needs research personnel and a flow of new scientific results. This in the long run should be a major reason for co-operating with universities. But is it the presence of universities that gives the incentive for co-operation or is it the demand from the industry that starts the exchange? It does not appear very likely that industry discovers from one day to the next that it has to co-operate with universities. It is also possible to argue that high technology industry growth precedes co-operation with universities (Braun and Macdonald 1982; Harrison 1982). This means that high technology growth started from a core of manufacturing industry applying R&D or from an industry that is shifting from traditional to high-tech industry. Examples are the radio and television industry, which transformed into the computer industry, and the basic chemical industry, which developed into the pharmaceutical industry. After the shift, co-operation between industry and university began to play a more important role, not because of their improved capacity to co-operate, but due to the birth of high technology firms.

C Networks and Network Formation

In the literature on technology transfer, university/industry relations are looked at in two different ways: technology push and market pull. An example of technology push takes place when science and applied science activities move from university to industry. Market pull means that business is forced to improve its products in order to open up new markets or expand market shares. In reality, neither technology push or market pull operate in isolation. On the contrary, they interact in a dynamic way (Massey, Quintas and Wield 1992). When emphasis is put on university as a motor for high technology growth, the perspective focuses on technology pull. This appears to be have been the dominant view of university and regional growth in recent years, at least in Sweden. The role of a regional industrial structure demanding university knowledge has generally played a minor role (Sörlin and Törnqvist 2000).

There are several reasons to assume that co-operation between university and manufacturing industry in the Stockholm region started as a result of an industry on the brink of change to a knowledge base. Several pharmaceutical and telecommunication firms were established in the region during the first decades of the twentieth century. There are few reasons, viewed from contemporary standard, to classify these as high technology firms. During the second half of the twentieth century, however, they began to make systematic use of research results in their innovation processes. Co-operation with and proximity to universities became the most important source of innovations. Since the 1980s, the electronics industry and information technology are two other new sectors that also draw fuel from university and industry co-operation.

In this section, we now look more closely at the innovation links between technology (i.e. specialisations in university institutes) and industry. In order to obtain an overview, the results from the survey of network linkages have been aggregated into 14 different industries and 13 technologies (see Table 4.24).

Table 4.24 Major innovation links between technology areas and industry in the Stockholm region (as percentages)

Industry/ Technology	01	02	03	04	05	06	07	08	09	10	11	12	13	Sum
Mining&Energy	2		2	2		1								6
Food&Drinks	1		1	1										3
Wood, Paper& Print	1	2							1					5
Chemicals	1	1			1			7	3	3				17
Rubber & Plastics				1										1
Basic Metals	1	2	1	2	1	1								7
Machinery	3	3	1	2	2	1					1	1		14
Electrical	3	2	1	2	2	2	1					1		14
Electronics	5	2			1	1	1			2			2	13
Construction	1	1	1											3
Transport& Telecom	2										1			3
Consulting			1		2	1		1	1		1			7
Prod. Services	2		2						1	1				5
Culture& Recreation	1												1	2
Technology Total	22	12	10	9	9	8	2	8	6	5	3	2	3	100

01 Information and Communications Technology
02 Production and Process Technology
03 Environmental Technology
04 Energy Technology
05 Materials Engineering
06 Automatic Control Technology
07 Microelectronics
08 Biotechnology and Pharmacology
09 Chemistry
10 Medicine
11 Transport Technology and Logistics
12 Air Transport and Space Technology
13 Culture and Media Technology

We find that out of 185 hypothetical links in the matrix, only in 63 do connections exist between manufacturing industry and technology. The most important technologies relate to information and communications, production and process activity, and activities concerned with the environment, which together comprise 44 percent of all links stated to be important. Co-operation is most frequent in technology areas related to chemical, electrical and machinery engineering which jointly account for 45 percent of the links.

The most intensively used links between manufacturing industry and technology account for more than three percent of all exchanges. They create seven major connections, which represent 27 percent of all university and industry co-operation in the Stockholm region. A closer look at the most intensive exchange displays a concentration to relatively few industries and technologies, see Fig. 4.21. They are the university-industry clusters operating within the metropolitan innovation system emerging from the empirical investigation. Information and communication technologies are most widely used and are present in machinery, electrical and computer software firms and electronics companies. The chemical industry has connections with three different technologies: biotechnology and pharmacology, medicine and chemistry. The machine industry has links with production and process technology as well as information and communications technology. Electrical engineering and microelectronics have one connection each to information and communications technology. The Stockholm cluster of innovation links between technology and manufacturing industry, which is illustrated below, accounts for more than a quarter of all university-industry links in the metropolitan region.

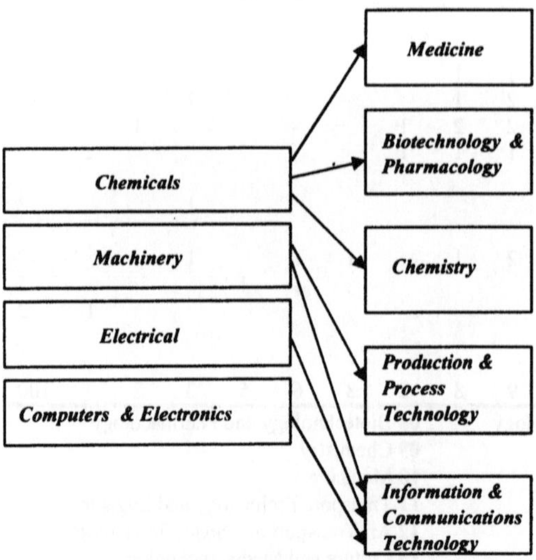

Fig. 4.21 The Stockholm cluster of innovation links between technology and manufacturing industry

A closer look at the industries in Fig. 4.21 shows that they contain the core of high technology industries. Biotechnology and electronics, especially, are important components in the group. The Swedish high technology industry is concentrated in the Stockholm region, and the concentration is especially strong in the counties of Stockholm and Uppsala. Estimations show that between 50 and 70 percent of high-tech firms, especially in pharmaceuticals and telecommunications, are located in these two counties (Axelsson 1992). This industry accounts for over 40 percent of all the R&D expenditure of private firms in Sweden.

Company size is another factor besides industry branch that has to be considered in order to understand the significance of the regional industrial structure for co-operation between university and industry. The survey indicated that large companies are far more likely to co-operate with universities in connection with innovation, as shown in Table 4.25. Moreover, the knowledge exchange is much more intensive. More than a third of companies with more than 500 employees have very intensive exchange with universities. In comparison, only four percent of small firms have intensive co-operation with universities. However size is probably not in itself the only decisive factor. There are numerous examples of regions dominated by large companies where co-operation with universities still remains at a low level. Probably it is more important for a region to have an industrial structure or product mix that demands exchange with universities than to be dominated by large companies.

Table 4.25 Intensity of university-industry co-operation and company size in the Stockholm metropolitan region 1997 (percentage of responses)

Company Size (Employees)	Never	Seldom	Intensive	Very Intensive	Total
1-19	91	5	0	4	100
20-99	41	23	29	7	100
100-499	41	15	29	15	100
Over 500	26	10	28	35	100

How can these results be interpreted from a wider perspective? High technology industry has several features that differentiate it from other types of business. It is characterised by frequent change of products, which means intensive innovation activity in early stages of the product cycles. These cycles are also much shorter, which presumably tends to make the industry more profitable. One of the most crucial factors for success in high technology is also accomplishment in R&D and the transformation of R&D into innovation. Today most industries depend to some extent on R&D, but this does not mean that all industries can be classified as high-tech. It is important to remember that knowledge creation is not a supporting production system as in traditional industry. R&D is probably the most important investment in this type of industry. From this perspective, it is evident that co-operation between universities and industries will be much more extensive in regions with a large high technology industry.

As mentioned above, high technology demands more R&D and is therefore more dependent on university co-operation. But what, more exactly, are the characteristics of high technology that promote university/industry co-operation? Is it possible to identify the factors that are particularly important? In order to understand better some of the mechanisms behind the co-operation between university and industry, the links have been divided into three different groups. The division is based on the three groups discerned in Table 4.25, which means that the 'intensive exchange' group represents links that has over three percent of the all links, 'medium intensive' links has two percent of all links, and 'low intensive' links has one percent or under of all industry and university co-operation. The composition of the different groups has been tested against other variables, which are given in Table 4.26. There are three variables that display a distinct pattern.

The group exhibiting high co-operation behaviour is correlated with the variable measuring the development of new products, and the groups representing medium and low co-operation show a significant correlation with the variable for improving new products. Finally, these last two groups show a significant correlation with the variable of production process improvements. The results indicate that strong co-operation links are required in the early stages of the product cycle. Later in the cycle, medium and low level co-operation links are significant for improving products and processes.

Table 4.26 High, medium and low levels of university-manufacturing industry co-operation correlated with different stages of the product cycle

Innovation Activity	High Co-operation	Medium Co-operation	Low Co-operation
Developing New Products	0.30[b]	0.17	0.23[a]
Improving New Products	0.13	0.24[b]	0.20[a]
Improving Production Processes	0.06	0.38[b]	0.37[b]

Notes: a Correlation is significant at the 0.05 level
b Correlation is significant at the 0.01 level

Viewed from this perspective, the extent of co-operation between university and industry will largely be determined by the regional industrial structure. If this observation is true it has several important implications. To provide regions with universities is not enough to create co-operation between university and industry. Another, and probably equally important factor, is the existence of a core of high-tech or traditional industry, which is growing into an R&D dependent industry.

D The Location of Innovation Co-operation Partners

As illustrated above, co-operation with other R&D institutes and with manufacturing firms takes place at different stages of the product cycle. When those contacts relate to innovation, they occur by definition in the first stages of the cycle. In the theoretical framework of metropolitan innovation systems developed earlier, we distinguished between the pre-competitive stage of co-operation and the competitive stage. In the first stage, the co-operation between partners takes place outside the competitive environment of market transactions. The interaction is less formalised and also involves contact with other R&D institutes (see Fig. 4.22) and firms working with the R&D institution in question, through participating in seminars and conferences, engaging in personal communication, and reading published materials. In this stage we also find co-operation under the auspices of research projects involving senior personnel, doctoral students or graduate students undertaking thesis work.

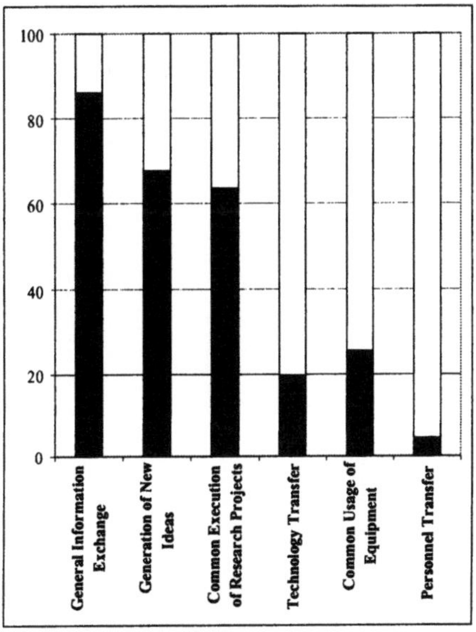

Fig. 4.22 Pattern of co-operation by R&D institutes in the Stockholm metropolitan region with other R&D institutes in 1997 (bars denote share of responding units stating strong interaction exists)

In the competitive stage, the co-operation becomes more formalised and contacts most often occur with other R&D institutes or firms (see Fig. 4.23) in the effort to perform development work aimed at market introduction of new products and services. This involves activities such as prototype development, pilot testing and finally market introduction. An observation worth making here is that there is a general tendency nowadays for firms to form strategic alliances with their competitors for joint work during the pre-competitive stage. After that, the product ideas developed are transformed into unique products within each co-operating firm. Examples are components in the car industry or in the telecommunications industry. For the transformation to function smoothly, it is important that the rules of interaction between universities and industries are clearly specified. The survey for the Stockholm metropolitan regions shows that both partners find these rules and cultures to be serious hindrances to the development of co-operation.

Fig. 4.22 and Fig. 4.23 show the prevalence of co-operation by R&D institutes in the Stockholm metropolitan region with other R&D institutions and with industrial manufacturing firms, respectively. The tables have been included to provide background to the presentation of results concerning the spatial extension of the co-operation networks. The first table shows that personnel transfer is not a frequent component of co-operation among R&D institutions in the Stockholm innovation system. Neither is it common to use laboratory equipment or borrow ideas directly. The co-operation among R&D institutions is concentrated on the exchange of ideas and undertaking of joint research projects. Two R&D institutes out of three are involved in research projects involving other R&D institutes.

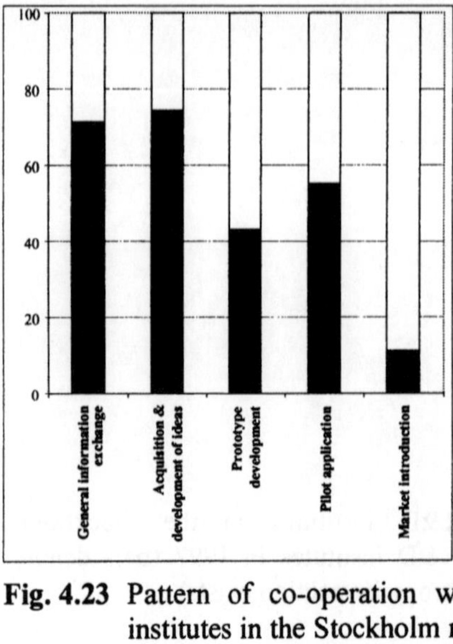

Fig. 4.23 Pattern of co-operation with manufacturing industry among R&D institutes in the Stockholm metropolitan region 1997 (bars denote share of responding units stating strong interaction exists)

Fig. 4.23 shows the prevalence of co-operation between university and industry. About one fifth of the R&D institutions have funding from external business sources in their research. It is natural for these institutions that they have involvement with business in their R&D activities. The table shows that during the early stages of the innovation cycle 25 percent of the R&D institutions surveyed were engaged in co-operation with business. We can also see that the level of co-operation declines markedly in the competitive stage of the cycle. About half of the R&D institutions state that they have strong interaction at the prototype development stage. Only one R&D institution out of five is engaged in co-operation with firms when new products and production processes are brought to the market.

Table 4.27 exhibits the spatial extension of the co-operation networks. The table applies to the sum co-operation activities between universities and industries. The difference in the spatial pattern is so small between university and business contacts that it is not meaningful to present them separately. The table shows that 44 percent of the partners are found in the Stockholm metropolitan region and another 22 percent in the rest of Sweden. One third of the contacts involve international partners. This is a considerable amount in view of the fact that these contacts are concentrated in the early stages of the innovation cycle. It means that they are strongly dependent on person-to-person contacts and is likely to generate a substantial amounts of travel and many short visits.

Table 4.27 Location of industrial and university innovation partners of R&D institutions in the Stockholm metropolitan region in 1997

All Networks	Metro Region	Rest of Sweden	Rest of World	Total
Architecture, Construction & Surveying	46	24	30	100
Biology, Chemistry & Medicine	38	19	42	100
Mathematics, Informatics & Physics	50	32	18	100
Electrical & Mechanical Engineering	46	19	35	100
Economics, Social & Geosciences	50	20	30	100
All R&D Institutions	44	22	34	100

Table 4.27 also reveals that the difference between academic knowledge areas is not very large when it comes to spatial patterns of networking. R&D institutions specialising in mathematics, informatics and physics depend to a larger extent on regional and national networks than the other areas. Institutions in the fields of biology, chemistry and medicine are more internationally oriented than the others. The conclusion of this analysis is that the metropolitan arena is very important for the co-operation networks. Almost every other contact occurs within the metropolitan region. The picture emerges of an activity which makes use of the positive externalities of the Stockholm metropolitan region, namely its transport and communications infrastructure.

Table 4.28 Relevance of previous contacts for co-operation of R&D institutes with manufacturing industry (number of contacts mentioned and percent of respondents stating involvement in contact type)

Contacts from the Time of	Metropolitan Region		Rest of Sweden		Abroad	
	no	%	no	%	no	%
University Studies	16	28	17	29	6	10
Doctoral & Postdoctoral Studies	24	41	24	41	15	26
Work in Research Institutes	28	48	31	53	22	38
Work in Business	15	26	18	31	8	14

Table 4.28 has been included as a further explanation of the patterns of co-operation discussed above. The table illustrates the relevance of early contacts of various types for the innovation co-operation between university and industry. Co-operation within the metropolitan region and within the rest of Sweden appears to depend largely on contacts established through previous work in research institutions and earlier academic studies, especially at the doctoral and postdoctoral level. Contacts abroad depend more strongly on prior work in research institutes.

The analysis has important implications for university policy. It shows clearly that contacts that have been established between university and industry during periods of earlier research work and doctoral studies tend to build up long-lasting networks which are later used for continued or extended co-operation in innovation work. The result reveals an active so-called 'third role' for universities, in the sense that university departments and individual researchers have a considerable long term benefit from belonging to an open network including other R&D institutes and research-oriented firms.

4.7 Concluding Remarks

The study has shown that metropolitan innovation potential is only partly determined by factors internal to the firms and institutions involved in the innovation activity. One of the external factors is hypothesised to be the nature of the innovation networks and other networks to which to a firm or institution belongs. According to this argument, the networks transfer positive externalities thus increasing the innovation potential, which in the longer run lead to enhanced economic growth through an increased frequency of innovation. The theory applies both to the manufacturing sector, which has traditionally been seen as the major carrier of innovation potential and to the service industries of the new economy, which are nowadays seen as major carriers of economic growth potential. The theory of national innovation systems considers that in this context it is mainly sector-specific factors which are important. The innovation potential therefore lies within a cluster of economic activity, which includes the firms, the labour they use, and the institutional structure in which they operate. Institutions differ across nations, and so does the competence given to labour through the educational system.

The metropolitan innovation systems approach argues that regional factors are necessary to explain the differences in the innovation potential among economic sectors. Economic development occurs in space, and the metropolitan regions are major actors in the economic development game. However, metropolitan regions differ in their capacity to act as incubators for innovation, see for instance Nijkamp (1990). These differences have been measured in the survey using a number of environmental indicators. One particular type of indicator involved asking the respondents to grade the qualities of the region according to a set of predetermined environmental factors. The factors are intended to measure the quality of supply factors in the metropolitan environment relating to capital, labour, manufacturing networks, manufacturing producer service networks, R&D networks, other innovation networks, infrastructure factors, and factors relating to public sector activities in the promotion of industrial development.

Fig. 4.24 provides a summary of the results of this analysis, focusing on the differences and similarities of firms and R&D institutions. The question addressed is how the different actors perceive one another as producers of positive externalities, how highly they value the quality of the urban infrastructure in the metropolitan region, both hard and soft, and how they assess the performance of the public sector. The intention is to provide guidance in the selection of policy to boost the potential of the region in serving as an efficient environment for innovation. The figure does not show how highly the respondents value the factors relative to one another, but how they rank the performance of the Stockholm metropolitan region as an environment for innovation in the selected aspects (a five degree scale has been used).

The most distinctive feature of the results presented in Fig. 4.24 is that industry and university perceive the quality of the environmental factors quite differently. For the R&D institutions, the most highly valued assets of the metropolitan region are factors relating to the supply of academic competence and the functioning of academic networks. They give a higher value to the hard and soft infrastructure factors in the region than manufacturing respondents. They are also somewhat more positive to the activities performed by public agencies in the promotion of university-industry linkages.

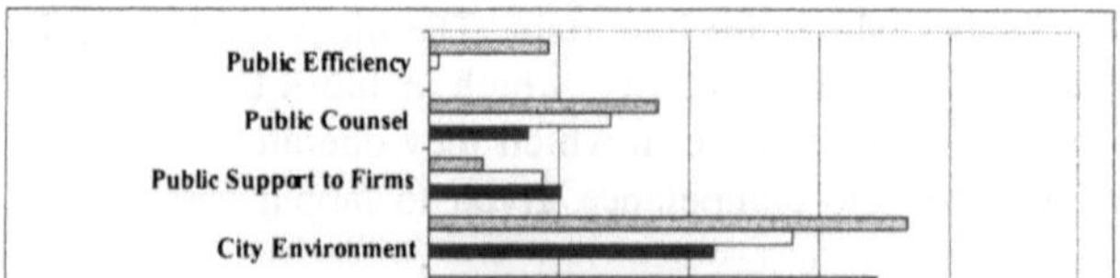

Fig. 4.24 Ranking a set of environmental factors in the Stockholm metropolitan region for innovation work in manufacturing firms, producer service firms and R&D institutes (average rank, using a five degree scale)

A second distinctive feature is that the average grading awarded is never higher than two in the five-degree scale. This indicates that the respondents did not seem any reason to praise the Stockholm region as an innovative environment to the extent that one might anticipate. Of course, the grading is relative and depends on expectations. It is however warranted to interpret the result as indicating that the respondents feel that the region is not performing to its full potential. A relevant benchmark would be to see the average being raised from around two to at least three. It should be noted that since the scale is relative and refers directly to expectations, this means that it is difficult to compare the results with other regions in Europe.

The third distinctive feature of the results is that the producer service firms grade the environmental factors more positively in several important respects than the manufacturing firms. Since the latter are more prevalent outside the core of the metropolitan region, this might be seen as an indication that the supply quality is not as well developed in the periphery as in the core. For the producer service firms, both the supply of customers and providers of intermediary inputs are highly graded. Apparently, the Stockholm metropolitan region provides ample opportunity for them to find new customers and choose appropriate suppliers of intermediate products.

A fourth feature is that all respondents give high grades to both the hard and the soft infrastructure factors. The variations are realistic in view of the differences between the respondents. The cultural factors are considered to be as well developed as transport and communications. It is also interesting that the respondents have given relatively low grades to the institutional infrastructure factors such as innovative networks and R&D capacity. One explanation might be that firms and institutions see these factors mainly as sectoral ones, operated and influenced mainly by national innovation actors rather than being a part of the metropolitan environment. Again, the results here show that there is a fairly distinct barrier between industry and university. Firms do not grade the quality of the contacts with universities as highly as contacts with customers and suppliers.

The fifth and final distinctive feature of the results presented in Figure 4.24 is the low rating by both industry and university of public activities relating to innovation. It should be noted that manufacturing firms are the least complimentary about the quality of public advice services, public support to firms, and public efficiency. This could be interpreted as a signal that the activities of the public sector are too national in character. Firms in the metropolitan region might not have enough experience of co-operating with public agencies, so the assessment is based on more general expectations.

The results of the comparison of the evaluation of environmental factors are important in that they further emphasise the results of other parts of the survey pointing at the lack of knowledge about other actors in the metropolitan region. The fact that one actor, for instance the R&D institutions, see their contacts in a different way from the firms with which they co-operate is also clear from responses to questions on co-operation in different stages of the innovation cycle. It is evident also from responses to questions concerning hindrances to co-operation between university and industry. Maybe one of the main contributions of a study such as this is to point out these differences in expectations. It is as if the actors in the region are not fully aware of the qualities that their co-operation activities have given to the region as an infrastructure system and incubator for urban innovation, see also Capello (2000).

Some Policy Implications

Sweden has a particularly strong statistical databases in comparison with other countries as regards population and employment. Housing data is generally regarded as more accurate in Sweden than in other countries. Industrial data exist at the regional level, although the sampling methods give accuracy only at a coarse regional level. On the other hand, regional economic accounts do not exist. These categories of socio-economic data are mentioned to give a picture of the regular flow of regional data through the Swedish planning system.

An important European research programme would be to create a joint policy for the regular provision of basic data at a detailed spatial level. It is also essential for comparative research on metropolitan innovation systems of the kind reported here that information to support policy indicators is collected for European regions on a common basis. We are only at the beginning of the important task of attempting to understand how the European urban system works. An essential component of any study using indicators is agreement on the definitions, collection methods, measurement periods, and imputation methods for spatial data. The ongoing work on the establishment of a Europe-wide strategy for the provision of spatial data is of central importance for such collaboration. The collaboration involves both work by government agencies in collecting, analysing and disseminating basic spatial data and work on designing cross-national research studies on impacts of policies at regional, national and international level.

The construction of indicators for regional development is a classical field of regional science research. Fundamentally, indicators can be seen as automated ways to assess the performance of policies, thus being part of the tool kit for policy evaluation. They can also be seen as probes to identify emergent regional problems, thus being instruments to provide new insights and perspectives in the selection of activity areas in need of public intervention. Swedish regional research has a tradition of providing government investigations with indicators of economic and social performance at the regional level. Accessibility indicators have been constructed on the basis of time-geographic research. In regional planning in the Stockholm area a number of useful indicators have been utilised to compare plan alternatives. New types of indicators have been used in conjunction with recent government investigations of sustainable transport in Sweden, where welfare indicators are employed to compare transport policy packages.

One important contribution of the current innovation survey project has been to create a platform for the identification of data needs for spatial policy, the selection of comparable definitions and spatial units, the selection of software platforms, organisational aspects, and questions of resource sharing. The Swedish view is that such a network could act as an innovative body in setting the stage for a truly spatial perspective in the European planning system. Among the challenges are the possibility of performing policy analyses for innovation potential, infrastructure networks, energy systems, pockets of renewable resources, and emerging threats to sensitive environments that cut across the national borders of Europe. Progress could be made in this field through examples developed within a network of research institutions.

If we regard the EU from an economic perspective, it can be argued that Europe has considerable problems compared to other regions of the world, such as the US or the Pacific Rim. Economic analysis reveals that labour mobility across state borders in the US is high, since people migrate readily to growth areas. Europe does not exhibit the same pattern of mobility because of barriers such as language and culture, but also because of historically established urban hierarchies. One question this raises is whether a mobility policy for economic growth would be at all possible in the EU.

The EU faces important decisions for the 21st century including the need to reduce the support for agriculture and switch it to urban areas. Such a shift of support is essential to keep people in (urban) places that are already, or threatening to become, areas of stagnation. A Europe-wide urban policy can play a role here, promoting the view of cities as engines for economic development. The challenge is to design territorial policies outside the agricultural area which favour balanced spatial development. The promotion of urban networks are at the heart of these policy ideas, and put the focus on policies that alleviate the pressure on urban areas, while at the same time allocating resources to growth-inducing urban economic activities.

There is also a need to carry out further empirical studies of innovation and location decisions among major European industries since many are faced with increasing movement of production factors. Such studies could be undertaken for service industries, i.e. industries that were formerly local, regional, and national. Examples could be taken from the consultancy sector, the 'culture' industry, and the transport and communications industries. The transport industry, for instance, is particularly important in view of the environmental impact resulting from an increase in interregional trade. These studies could also encompass the impact of increased competition on the location of intermodal facilities, for instance, airports. The integration of Europe coupled with a deregulation of the airline industry, and the establishment of a private airport industry, will fundamentally change the competitive situation among European regions.

The changes also interact fundamentally with the development of the metropolitan innovation system. In the long-term, it is likely that the manufacturing industries of Europe will be looking for both economies of scale and scope, and will choose locations that reduce their need for transport and communications given the situation where governments are emphasising the need to reduce transport requirements. This may give rise to substantial location changes not only due to the relocation of existing capacity, but also because of emergent differences in profitability.

The other important factor for the European urban system is the impact of deregulation and of rapid technical development in the service industries. These processes may have substantial effects on labour demand, and also on location developments. The long-term question is what location patterns will emerge at the European level for those sectors that have until now been protected. A particularly important example is provided by the higher education sector, both because of its market potential and its stated importance as an engine for regional development. Another important example is the financial sector that has for the moment a dual development pattern. It is developing locally in order to provide customers with more complex offer of services, and internationally to secure risk capital and fund large-scale investment projects. In the context described above, it is hoped that the research work reported here on metropolitan innovation potential and networks will make a contribution to furthering understanding on how metropolitan regions work as innovation systems in a regional, national and international context.

5 A Retrospect

It is customary for a concluding chapter to summarise the findings presented in the preceding part of the book. In this case it is a somewhat daunting task. The three metropolitan region studies that make up the heart of this volume have covered an enormous amount of material. A great effort has been made to provide consistent and comparable data for three selected metropolitan regions (Barcelona, Stockholm and Vienna). The project has in fact produced a unique resource base for the analysis of metropolitan innovation activities and potential. It differs from previous regional innovation studies in a number of ways.

Firstly, the project involved three types of survey: surveys of manufacturing firms and surveys of two types of organisations providing assistance or support to manufacturing units for the development and/or introduction of new/improved products or processes.

Secondly, the survey methodology involved making postal surveys of local manufacturing units, producer service providers and research institutions. These were designed to elicit basic quantitative data concerning the pattern and scale of innovation activities within the three regions. In this way we were able to build up a comprehensive picture of the innovation and network activities in each of the three study regions. Much of the data collected has been displayed in the form of tables summarised in the Appendix.

Thirdly, the focus was on the activities of the local unit rather than the enterprise of which it was part, in order to identify innovation activities and capabilities at the metropolitan level.

Fourth, the operation of the project activities were optimised in various ways, including the use of professional translation in the production of the questionnaire schedules, the exercising of central control over the sampling methodologies, the production of the questionnaires, and the allocation of sufficient resources both to enable large samples of local units to be surveyed, and to allow sufficient time for the surveys to be conducted.

To summarise the outcome is not easy, but it is possible to try to assess the extent to which the project succeeded in achieving the goals that were foremost in the minds of the research partners. This chapter offers such a retrospective, and attempts to identify the policy implications if Europe's regions are to meet the new challenges brought about by the emergence of globalise knowledge-based economy.

The principal objectives of the study were laid out in some detail in Chapter 1, thus there is no need to repeat them in detail. The first objective was, simply, to describe, compare and try to understand the similarities and differences in innovation systems of three selected metropolitan regions. Until now, studies on regional innovation systems had been rare, with international comparisons left mostly implicit. To produce consistent and comparable data from the perspective of the systems of innovation approach was the second major objective of this project. A third objective of the study was to throw some light on a complex series of issues including innovation and networking activities, economic performance and regional development.

From another point of view the project as a whole could be considered as a test of the set of intellectual commitments that guided the work. In particular, the basic notion that oriented the project – the 'metropolitan innovation system' – is a complex and somewhat problematic one. Our concept of innovation was rather broad and not necessarily tied to leadership in a given technology, but to effective competitive performance in a dynamic context. In addition, the authors of this book are well aware that innovation systems are not contained neatly within national and regional borders. The trans-regional and trans-national aspects of technological change, and of the key actors involved, have grown increasingly prominent in recent years. Nonetheless, we believe that localities (regions and countries) continue to be meaningful units of observation and policy.

5.1 Metropolitan Differences and Similarities

Certainly the broad view of technological innovation that we set out in Chapter 1 and that has guided this study implies certain commonalities. It applies to metropolitan economies in which profit-oriented firms are the principal providers of goods and services, and where central planning and control are weak. These conditions hold in all three metropolitan economies examined, although in some cases (such as the Vienna metropolitan region), part of the economy is still nationalised, and regional and national governments do attempt to mould the shape of economic development in a few economic sectors.

Innovation involves much more than R&D, and the set of formal and informal institutions that influence the innovation capabilities of a metropolitan region and how these are enhanced extend far beyond those that directly impinge on technological innovation. The character and effectiveness of a metropolitan system of schooling, training and retraining strongly determine the supply of skills. A diversified education system, including well-established vocational training facilities with the capacity to prepare a large number of highly skilled people in different fields, is a major asset possessed by both the metropolitan regions of Vienna and Stockholm.

Education not only determines the supply of skills, but also influences the attitude of workers towards technological advance. So too does the set of institutional arrangements that lead toward different kinds of inter- and intra-firm co-ordination. The metropolitan economies of Stockholm and Vienna are characterised by a system that favours long-term employment and closer relations with the workforce. This makes it easier to pursue strategies based on incremental process or product innovation of the sort which builds on the knowledge embodied in the labour force. The presence of institutions that enable to co-ordinate vocational training and pressure from powerful trade unions privileges forms of production based or high-cost and highly-skilled labour focusing on high value-added product lines. But such an institutional set-up tends to militate against radical innovation in new product lines, since it is rather difficult to recruit expertise from outside the firm and to change the skill categories of the workforce.

In all three regions, education including university education is provided in public institutions. Governments also have major responsibility for funding basic research, but there are major differences regarding how much they do so. This is evident by comparing the R&D-intensity as percentage of GNP and the scale of defence-related R&D out of total budget appropriation for R&D in 1991: Sweden 2.5 and 24.7, Austria 1.5 and 0.0, Spain 0.9 and 18.4, respectively. The share of military expenditure has been in rapid decline for several years. Nevertheless, the Swedish and Spanish governments still finance a significant part of military R&D performed by firms in the metropolitan regions. There are also differences across the regions regarding how much basic research they do, and where basic research is carried out, as well as in the kind of applied R&D that the governments finance. But the divergences across our set of metropolitan regions have to be understood as differences between creatures belonging to the same species. If Eastern European metropolitan regions such as Budapest, Prague, Warsaw, St Petersburg, or Moscow had been included, things would have been rather different.

Within our three metropolitan regions, the differences in the innovation systems seem largely to reflect differences in economic and political circumstances and priorities. The lower income metropolitan region, Barcelona, tends to differ from the higher income ones, Vienna and Stockholm, in the kinds of economic activities where it can achieve comparative advantages and in the internal demand patterns. These differences shape the nature of the technological change.

Chapters 2-4 provide evidence about the nature of three metropolitan innovation systems. One fact to emerge is that the past of a metropolitan innovation system influences its present performance. This is an example of what is more generally called 'path dependence'. It is also clear that metropolitan innovation systems show a very high degree of regional and national specificity. Although systems may achieve similar targets in terms of growth of output, productivity, exports, and so on, they achieve them by means of an institutional set-up that is country and even region specific. The three innovation systems have all taken shape in countries with advanced economic performance. This obviously rules out any question of classifying them as either 'good' or 'bad' systems. To rank them in order of efficiency or performance would not only be technically difficult and fraught with theoretical difficulties and intellectual dangers, but would probably not be legitimate, since the regions do not all face the same challenges. They differ in their size, industrial and technological specialisation, and scientific and technical potential. From this point of view, the concept of 'efficiency' of a spatial innovation system has inherently a specifically territorial dimension.

5.2 What Does the Study Tell Us?

Several interlinked conclusions can be drawn from the account of recent developments in the three metropolitan innovation systems. However, before summarising these, it is important to restate two key factors. These set the parameters within which these changes occur and remind us of a third generic element that constitutes a key response (see Cooke 1996).

- *First*, globalisation, especially in the sense of an increased capacity for the interpenetration of domestic markets by foreign producers, is a key feature of the present economic situation. Together with the deregulation of markets, it is creating enormous uncertainty, putting firms in a situation where doing anything is better than doing nothing (Cooke 1996). This creates a mentality which Sabel (1994) has referred to as 'bootstrapping reform' among firms, unions and regional governments.

- A *second* key parameter of the present turbulent context is the sense that innovation is the key to survival. By innovation is meant not only product or process innovation but also managerial, organisational and even cultural innovation. This implies that change may not be sufficient. Firms must engage in heroic acts to reinvent products, processes and culture not once but continually in order to sustain competitive advantages. Some contexts are better than others in which to achieve this.

- *Third*, networking is the buzzword usually given to this form of interactive exchange, learning, trust building, partnering and empowering practice. If indeed the metropolitan regions of Stockholm and Vienna do have a comparative competitive advantage over the Barcelona region, it lies essentially in the rich milieu of firms, institutions and informed personnel which constitute their metropolitan economies. Through the information flows made possible by the innovative regional business infrastructure found in the regions, firms are able to provide a rapid response to the turbulent conditions. Even if the institutional context in which the public and private actors in the innovation systems find themselves is out of synchronisation with business needs, there is enough useful information flowing informally for firms to be able to experiment with their own *bootstrapping* solutions. These then may become demonstration models for the rest.

The following are the main findings and conclusions that can be derived from Chapters 2 to 4:

- *First*, there is a clear trend towards higher knowledge intensity in all economic sectors, and indications that higher knowledge intensity leads to better performance of the firm at sectoral and regional levels. This is particularly clear in the case of Stockholm, where the subdivision of the manufacturing sector into segments with different levels of knowledge intensity makes it possible to explicitly demonstrate the high correlation with positive innovation indicators. Manufacturing firms tend to be good at improving their existing product lines and building incremental innovations into their products, sometimes tied to changes in production technology and new organisational practices. In general, however, these innovation activities are based on product and/or process innovation of the kind that builds on knowledge embedded in the labour force, rather than on radical innovations in new product lines.

- *Second*, the results show a number of differences in innovation behaviour. Generally larger firms are more innovative than smaller firms. They are also more intensive users of wider innovation networks including international and even global ones. There are exceptions such as high-tech spin-off firms that have global interactions, but these are a minority in the metropolitan regions considered. In traditional low-tech sectors, significant productivity gains may be achieved through innovation with respect to production, organisation, logistics and marketing. In appropriate circumstances, innovations of this kind may be highly significant. However, the extent to which practical knowledge, i.e. know-how, actually informs innovation processes reflects the internal organisation and management style of the firm.

- *Third*, R&D expenditure and the presence of research staff are more prevalent in larger than in smaller firms and in high-tech than in low-tech firms. Even though innovation is still very much an internal process, innovativeness is not necessarily closely linked to the presence of continuous on-site R&D activities. In contrast, the study clearly illustrates that learning-by-interaction (i.e. learning-by-doing by communication and exchange with others) is important to the development of the innovation capacity. For example, in Vienna about 50 percent of the respondent manufacturing firms which had developed product innovations did so in collaboration with other organisations. This nevertheless implies that the remaining half of the firms involved in product innovations, operated in isolation. The knowledge acquired from other organisations was generally combined with that development through learning inside the firms. This clearly indicates that the innovation process is far more complex and diverse than suggested by the stereotype of linear progression from basic scientific research, to applied R&D and to the application of innovations in production.

- *Fourth*, innovation takes place most effectively in an institutional environment where learning is fostered through intensive information exchange between organisations and where the process of developing trusting relationships (expectations of honest, non-opportunistic behaviour) between organisations is facilitated by the institutional set-up. Aspects such as legal systems, norms and values differ widely between the metropolitan region of Barcelona and the other two regions. In Stockholm and Vienna, an institutional structure emphasising consensus and long-term commitments has given workers the abilities and competence to increase the value of existing technologies through incremental and continuous innovation activities in products and processes. It is not only individual, but especially organisational learning – both within firms and between firms and other organisations – that provides the key to the higher level of innovative capacity characteristic for these two metropolitan innovation systems.

- *Fifth*, innovation capability is not distributed evenly between organisations. There are clear sectoral differences supporting the findings of Pavitt (1984). Firms in supply-dominated sectors such as clothing and furniture generate few important innovations themselves, but tend to import them from other firms. There are also scale-intensive sectors, such as the food industry, in which process innovations predominate. Moreover, science-based producers, such as firms in chemicals and electronics, tend to develop both new products and processes, often in close collaboration with universities and research institutions. Finally, there are specialised suppliers such as instruments, engineering and computer software firms characterised by frequent product innovations, frequently developed in close collaboration with their customers. It is the third sectoral pattern that has tended to dominate consideration of innovation processes and that conforms most closely to the incorrect view that innovation is confined to the high-tech industries.

- *Sixth*, the sharing of knowledge via learning-by-interaction may take a variety of forms. It may involve the acquisition of existing knowledge from other organisations, such as firms, universities, and research institutions, through the purchase of technological licenses. Or it may involve other less formal types of exchange that result from routine interactions with suppliers and customers. Once acquired, such existing knowledge may be integrated with other knowledge elements in new combinations to produce innovations of various kinds. The shortening of hierarchies, the decentralisation of responsibility to lower-level employees and the flexibility of work organisation may be seen as key elements in creating an environment in which learning-by-interacting may prosper (Edquist and Rees 2000).

- *Seventh,* in the metropolitan regions investigated in our study, learning-by-interaction takes place predominantly in the pre-competitive rather than the competitive stage of the innovation process, independent of the type of networking involved. External information tends to be particularly relevant during the early phase of the innovation process, when the perception of problems and evaluation of technological possibilities is important. This points to a problem not fully addressed in the metropolitan regions, viz. the co-operation within university and industry in the shift from co-operation to competition. It is evident from the case studies that the two systems have very different cultures which stimulate one another, while at the same time causing management problems on both sides. In these dynamic contexts, the public support institutions – mainly conceived as part of the national innovation systems model rather than the regional one – often seem to find themselves in a problem-solving rather than opportunity-enabling position.

- *Eighth*, extremely important for the innovation potential are the relations found along the value chain (customers, manufacturing and producer service suppliers). These characteristically go beyond simple market relations, and basically involve two types of interaction. One is interdependent, functioning as a co-operative mode relying on tacit performance agreements, trust and reciprocal adjustments. The second is more of a contractual, competitive or arm's length mode, where interfirm trust and familiarity may be very limited. Both types of transactions appear to coexist in the metropolitan innovation systems. Much less frequently, organisations collaborate with others in order to produce new knowledge, and potentially innovations, directly. Occasionally, firms and research institutions in the Stockholm and Vienna regions have formed R&D consortia with the objective of generating new knowledge as the basis for innovation. But in general, industry-university linkages have been identified as one of the weaker elements of the metropolitan innovation systems concerned. Although scientific research in the metropolitan regions of Vienna and Stockholm is up to world standard, it seems generally that the universities have difficulty in transferring research results into commercially viable innovations and new products. This clearly reflects mismatches between the needs of private firms and the science and research infrastructure, which remains under-utilised except when successful programmes of public-private partnership catalyse more productive relationships.

- *Ninth*, there is clear empirical evidence that interactive learning, which takes place within networks of organisations, is localised and territorially specific. One of the reasons for this is that so much of the synergy is transmitted through the innovation system via personal contacts established earlier in the career of labour, again demonstrating the path-dependency of innovation potential. This phenomenon is essentially based on the observation that knowledge creation and learning (in the sense of integration of knowledge) is context-dependent or – more specifically – embedded in socio-economic networks that rely on close interaction and the exchange of tacit knowledge. The Vienna case study, for example, reveals that one in three manufacturing firms involved in collaboration with other organisations related to product innovation had at least one partner located within the metropolitan region. The marginal cost of transmitting knowledge (in contrast to information), and especially tacit knowledge, rises with distance (Audretsch 1999). Tacit knowledge is best transmitted via face-to-face interaction and through frequent and repeated contact. But it is also possible that effective collaboration and interaction may take place within networks that are highly dispersed. The growing sophistication of information and communication technologies opens up new possibilities for the development of effective networks of organisations based upon spatially dispersed interaction.

- *Tenth*, firms with networking activities tend to maintain linkages at more than one spatial scale (regional, national, European, or global). But there are differences between types of firms. Companies belonging to larger foreign owned corporations are – due to their corporate links – usually more integrated within European and global networks. Moreover, innovative firms tend to be better integrated into networks than non-innovative firms. Much has been written about the role of economic clusters in the recent literature on national innovation systems. In many of these studies the regional dimension is completely suppressed. It emerges clearly from our three metropolitan studies that it is both scientifically fruitful and highly policy-relevant to know how these clusters operate in concrete regional settings. Without such knowledge, it is easy for national policy attempting to materialise the innovation potential of an economy to go astray. It has been shown that economic clusters are regionally based, with metropolitan regions as their hubs. Since the complex interaction between university, industry, and public policy unravels at the regional level, economic growth policy needs to be implemented, to a much larger extent than before, at the metropolitan level. This provides challenges for both national and European policy-makers who have traditionally felt rather uncomfortable about handing over power to the regions.

- *Finally*, it is important to note that flows of tacit knowledge through informal interactions and the mobility of personnel have an important positive effect on innovation performance, especially on the ability of firms to detect, adapt, adopt and use new knowledge. International flows of knowledge are still primarily embodied in traded equipment, but the internationalisation of innovation networks is accelerating, and expanding the range of types of knowledge exchange and modes of knowledge transmission. In this context it is interesting to note the substantial importance given by the respondents in our survey to fairs, exhibitions, and conferences as a nexus for knowledge exchange and contact-creation. It is also reflected in the emergence of new producer service activities in the form of science brokers, science journalists, trade fair catalysts, and congress organisers at the metropolitan level.

5.3 Policy Implications

The emergence of an innovation-led global economy, with all its uncertainties and the difficulties of national macro-economic management, demonstrates that national and regional governments need to reconsider their role and how their actions may affect the competitive advantages of firms in world markets. While globalisation will require governments to surrender some of their traditional tasks to market forces, those which remain are no less vital to the economic welfare. On the one hand, an increasing role is being recognised for subnational levels of governance and, on the other, new or strengthened supranational governance institutions may be necessary.

In Chapter 1 we have characterised the systems of innovation approach as a conceptual framework for gaining a deeper understanding of metropolitan systems of innovation. On this basis we can identify two types of policy implication: general implications that derive from the characteristics of the approach, and specific policy issues which derive from the empirical studies described in Chapters 2 to 4.

The general policy implications are related to the innovation subsystems and especially the institutional sector, the relations between these subsystems, lock-in situations and demand side instruments (see Edquist and Rees 2000). One important implication derives from the fact that innovation is interactive, consequently innovation policy should not only focus on the actors of the system, but also the relations between them. Fruitful interactions between the different components and actors of a metropolitan innovation system can be hampered by a number of factors (see, e.g., Guinet 1997):

- Conflicting incentive structures [cultures] of co-operation partners, e.g. university versus industry;
- Market failures that may hinder access by – especially smaller – firms to information, technology and know-how, or may weaken firms' incentives to invest in technology or absorptive capacity [e.g. high transaction costs];
- Co-ordination failures whereby agents and actors fail to recognize their complementary needs and assets;
- Lack of managerial competences reducing the understanding of the role of technology and innovation in the competitive strategy, and of the benefits of collaborative strategy of knowledge acquisition and dissemination;
- Financial markets unable to assess firms' investments in interactive learning, e.g. underdevelopment of specialised venture capital;
- Last but not least, lack of human resources on which to build capabilities to absorb external knowledge.

Following Guinet (1997) policy responses to these market and systemic failures can be categorised as follows:

- *Regulatory framework conditions conducive to innovation*, i.e. macroeconomic stability to assist strategic investment decisions, labour market policy to facilitate mobility of science and technology personnel, competition policy to increase propensity to innovate while allowing co-operative behaviour in building and utilizing the knowledge base;
- *Infrastructure policies* to fill gaps in the knowledge base, e.g. support to basic research or to the development of infrastructure technologies, and to make the public infrastructure in general and the science and research sector in particular more responsive to evolving business needs;

- *Catalytic actions to stimulate inter-firm co-operation* and public/private partnerships in knowledge generation in general and the development of generic technologies in particular;
- *Actions to improve the environment for high technology start-ups* and for starting up and developing innovative businesses in general;
- *Technology diffusion policies* to correct market failures in knowledge transactions due to supply or demand factors.

National and regional governments will have to reconsider their role. In doing so governments should become aware of the need to envisage their efficiency in the broader framework of an overall policy strategy to facilitate flows of knowledge and other interactions between the different components of an innovation system. Hereby the following trends and principles should be taken into account:

- *First*, there is an overall shift from traditional supply-side diffusion measures towards policies, which reflect a more interactive model of innovation and recognize innovation and diffusion as interdependent processes as discussed in Chapter 1. This translates into emphasizing a greater role for demand-driven policy programmes, network-building initiatives, and policy measures to upgrade the technology diffusion infrastructure and improve its relevance for and accessibility by smaller businesses.

- *Second,* the nature of the knowledge being diffused is changing. Greater emphasis has to be laid on diffusing soft organisational technologies of both technical and quality management [e.g. ISO certification] as well as on hard technology of information and communication management [e.g. Internet and e-commerce].

- *Third,* severe constraints on public finances have robbed governments of the means even if they still have the will to engage in large-scale and widespread interventions. Like firms governments face a difficult learning process nowadays as they work out to pursue the public interest efficiently and effectively in an increasingly global and increasingly knowledge-based economy. Governments cannot leave entirely to the discretion of firms and other organisations the choice of organisational adjustment processes in the face of globalisation and rapid technological change. They have to remain powerful framers of the business environment even in a more global knowledge-based economy. They will have to make choices between market-reliant, market-regulating, market-supplementing and market-displacing strategies. A redesign of the environment for business is necessary as foreign direct investment and the Internet economy become more and more important features of the global economy, and as the knowledge intensity of internationally traded goods and services increases.

- *Fourth,* there is a need for policy co-ordination, to rank order the main policy objectives, build the overall innovative capacity of firms, and stimulate the adoption of specific technologies. This is crucial for achieving greater coherence between programmes with different targets (specific technologies, sectors, institutions or regions) and offering different diffusion services as technical assistance, information and training.

- *Fifth,* in order to arrive at maximum leverage, technology policies should build on existing relationships in innovation systems, in particular networks within which firms collaborate and exchange tacit knowledge.

Technology diffusion policies in the three regions mostly address the needs of the highly heterogeneous population of small and medium sized enterprises, from high technology start-ups to non-innovative firms in mature industries or conventional services. This point can be illustrated by considering the role of technology diffusion policies in the promotion of new technology based firms, which has become a priority objective in most advanced economies. In Austria, for example, there exist three related programmes: the Seed-Financing Programme and Technology Marketing Austria TecMa, both in close cooperation and the Young Entrepreneur's Programme by Bürges. While the Seed-Financing Programme and Bürges concentrate on financing, TecMa provides consultation in connection with the commercial exploitation of research results and inventions. Financing is also closely related to other innovation financing schemes such as the so-called FFF general programme and the ERP Small and Medium Sized Technology Programme. Equity capital guarantees by the FFF are also interesting to technology-oriented start-ups.

In general and much more than in the Austrian case technology diffusion policies have to be tailored to the particular needs of new technology-based firms. Hereby, governments have to take into account two overriding considerations:

- *First*, in many activities technology based firms do not generate themselves novel knowledge through formal R&D activities, but rather test on the market new ways of combining existing technical solutions. In other words, such firms are knowledge-intensive but not always R&D intensive. They need to get easy access to external sources of ideas and talents. Spin-offs involving technology, personnel and business opportunities from the R&D efforts of large corporations are important in this respect. In addition, the more diversified is the research portfolio of the science and research sector in a region the more it is a good springboard for new technology-based firms in a wide spectrum of activities.

- *Second*, new technology-based firms agglomerate within regional clusters of knowledge-intensive activities. This spatial dimension makes more important the choice and co-ordination between different levels of co-ordination and implementation of technology diffusion policy measures on the one side and between the nation-state and the region-state on the other.

The efficiency of technology diffusion policies in promoting new technology-based firms depends on a number of conditions subject to government influence. Regulation is necessary, but over-regulation hinders the development of enterprises, innovative enterprises in particular. New technology-based firms have already to cope with an exceptional level of technical and commercial risks. Thus, they tend to be more vulnerable than other enterprises to the additional uncertainties that government action may create in their tax, regulatory or macroeconomic environment. Rewards expected by the entrepreneurs and their financiers should be proportionate to the risk they take. Consequently, framework conditions should not affect unfavourably on this risk versus reward ratio [for example, tax system discriminating against capital gains, high interest rates], but should provide mechanisms for rewarding investment in new technology-based firms, e.g. via efficient secondary stock markets and novel channels for providing risk capital to the innovation system. An example in line with these considerations is the reorganisation of the national institutions for the promotion of the innovation system in Sweden. A new government agency being set up in 2001 has the exclusive mission to finance R&D and related activities in support of the Swedish innovation system.

In summary, empirical analyses are absolutely essential for the design of specific innovation policies. It is also important to be able to compare national and regional systems of innovation in a systematic and detailed way. The systems of innovation approach is a conceptual framework highly suited to such analyses. It is attractive for this purpose because it places innovation at the very centre of focus and is able to capture the differences between systems.

APPENDIX A

The Vienna Metropolitan Innovation System

Manufacturing Sector:	Tables A.1 – A.12
Producer Service Sector:	Tables A.13 – A.20
Science & Research Sector:	Tables A.21 – A.28

Table A.1 Selected characteristics of manufacturing sample firms (1996)

	Sample Firms with					
	Regular R&D[b]		No R&D[c]		Total	
	no.	%	no.	%	no.	%
Corporate Status						
Single Establishment	43	44.8	67	68.4	110	56.7
Multi-Establishment	53	55.2	31	31.6	84	43.3
Main Plant	*30*	*31.3*	*15*	*15.3*	*45*	*23.2*
Branch Plant	*23*	*23.9*	*16*	*16.3*	*39*	*20.1*
Total	96	100.0	98	100.0	194	100.0
Employment Size						
≤ 49	28	29.5	57	60.0	85	44.7
50 – 99	22	23.2	22	23.2	44	23.2
100 – 499	35	36.8	15	15.7	50	26.3
≥ 500	10	10.5	1	1.1	11	5.8
Total	95	100	95	100	190	100
R&D Expenditure [a]						
None	0	0.0	35	43.7	35	21.9
0.1 – 3.4	35	43.8	32	40.0	67	41.8
3.5 – 7.9	21	26.2	7	8.8	28	17.5
≥ 8	24	30.0	6	7.5	30	18.8
Total	80	100	80	100	160	100

Notes: a annual average of 1994-1996
b regular R&D firms are permanently engaged in research or development process
c no R&D firms are only occasionally or never engaged in research or development process

Table A.2 R&D of manufacturing firms (1996) disaggregated by size classes

	Firm Size by Employees									
	≤ 49		50 – 99		100 – 499		≥ 500		Total	
	no.	%[b]	no.	%[b]	no.	%[b]	no.	%[b]	no.	%[b]
***R&D Expenses* (in % of turnover)**										
No R&D	21	31.4	8	21.1	6	13.3	0	0.0	35	21.8
0.1 - 3.4	26	38.8	14	36.7	23	51.1	4	40.0	67	41.9
3.5 - 7.9	10	14.9	8	21.1	8	17.8	2	20.0	28	17.5
≥ 8	10	14.9	8	21.1	8	17.8	4	40.0	30	18.8
Total	67	100	38	100	45	100	10	100	160	100

Notes: a average 1994-1996
b percentage figure as column %

Table A.3 Innovativeness of manufacturing firms (1996) by age

	Year of Founding /Establishment									
	Before 1969		1970-1979		1980-1989		Since 1990		All	
	no.	%[a]	no.	%[a]	no.	%[a]	no.	%[a]	no.	%[a]
Firms										
Non-Innovative	36	53.7	11	16.4	11	16.4	9	13.5	67	100
Innovative[b]	78	59.1	21	15.9	14	10.6	19	14.4	132	100
Total	114	57.2	32	16.1	25	12.6	28	14.1	199	100

Notes: a percentage figure as row %
b firms with product innovations in 1994-1996

Table A.4 Manufacturing firms with product innovations (1994 - 1996)

	Corporate Status							
	Single Establishment		Main Plant		Branch Plant		Total	
	no.	%[a]	no.	%[a]	no.	%[a]	no.	%[a]
None	41	59.4	11	15.9	17	24.7	69	100
Further Development of Products[b]	49	53.8	24	26.4	18	19.8	91	100
New Products[c]	16	66.6	4	16.7	4	16.7	24	100
Total	106	57.6	39	21.2	39	21.2	184	100

Notes: a percentage figure as row %
b more than 50% of the products were substantially developed further during the years 1994 to 1996
c more than 50% of the were newly introduced in the product programme 1994 to 1996

Table A.5 Sources of external information for product and process innovation (1996)

	Product Innovation Activities		Process Innovation Activities	
	no.	%[a]	no.	%[b]
Sources				
Customers	123	93.9	39	36.4
Suppliers	65	49.2	66	61.7
Competitors	79	59.8	44	41.1
Research Institutions	29	22.0	25	23.4
Producer Services	27	20.5	44	41.1
Fairs/Exhibitions	83	62.9	50	46.7
Specialised Literature	79	59.8	68	63.6
Media	33	25.0	14	13.1
Internet	6	4.5	5	4.7
n	132		107	

Notes: a percentage of all firms with product innovation activities
b percentage of all firms with process innovation activities

Table A.6 Motives for exercising network activities (1996)

	Network Activities with					
	Research Institutions		Producer Service Providers		Other Manufacturing Firms[b]	
	no.	%[a]	no.	%[a]	no.	%[a]
Motives						
Risks/Cost Reduction	6	13.3	40	38.1	50	64.9
New Technological Opportunities	26	57.7	17	20.5	36	46.8
Know-How-Takeover	15	33.3	37	35.6	33	42.9
Financial Resources	1	2.2	21	25.3	15	19.5
Funding Requirements	22	48.9	30	36.1	6	7.8
n	45		83		77	

Notes: a percentage of manufacturing firms cooperating with the corresponding network partner
b relation of manufacturing firms with manufacturing suppliers, customers or competitors

Table A.7 Problems with exercising network activities (1996)

	Network Activities with					
	Research Institutions		Producer Service Providers		Other Manufacturing Firms[b]	
	no.	%[a]	no.	%[a]	no.	%[a]
Problems						
Problem with Project Management	5	21.7	11	20.8	4	8.7
Budgeted Cost Overrun	9	39.1	20	37.8	6	13.0
Unintended Knowledge Drain	3	13.0	9	17.0	22	47.8
Coordination Difficult	6	26.1	12	22.7	15	32.6
Different Capability	7	30.4	18	34.0	16	34.8
Confidential Relation/ Secrecy	4	17.4	4	7.5	18	39.1
Loss of Independence	1	4.3	6	11.3	4	8.7
Lack of Schedule Effectiveness	11	47.8	16	30.2	13	28.3
n	23		53		46	

Notes: a percentage of manufacturing firms cooperating with the corresponding network partner
b relation of manufacturing firms with manufacturing suppliers, customers or competitors

Table A.8 Network activities of manufacturing firms with customers (1996)

	Metro-politan Scale		National Scale		European Union		Central / Eastern Europe		Mondial Scale		Total	
	no.	%[b]	no.	%[b]	no.	%[b]	no.	%[b]	no.	%[b]	no.	%[b]
Pre-Competitive Stage												
Information Exchange												
Strong Ties[a]	40	30	38	29	35	27	14	11	5	4	132	100
		(40)		(38)		(35)		(33)		(24)		(36)
Weak Ties[a]	4	16	6	24	7	28	3	12	5	20	25	100
		(13)		(19)		(26)		(33)		(56)		(23)
Identification of New Ideas												
Strong Ties[a]	33	27	35	28	34	28	13	11	8	7	123	100
		(33)		(35)		(34)		(31)		(38)		(34)
Weak Ties[a]	11	32	9	26	8	24	4	12	2	6	34	100
		(35)		(29)		(30)		(44)		(22)		(32)
R&D												
Strong Ties[a]	28	26	28	26	30	28	15	14	8	7	109	100
		(28)		(28)		(30)		(36)		(38)		(30)
Weak Ties[a]	16	33	16	33	12	25	2	4	2	4	48	100
		(52)		(52)		(44)		(22)		(22)		(45)
Competitive Stage												
Prototype Development												
Strong Ties[a]	25	26	26	27	29	30	12	12	5	5	97	100
		(31)		(33)		(32)		(30)		(23)		(31)
Weak Ties[a]	19	32	18	30	13	22	5	8	5	8	60	100
		(37)		(35)		(36)		(45)		(63)		(38)
Pilot Projects												
Strong Ties[a]	29	26	29	26	32	29	14	13	8	7	112	100
		(36)		(36)		(36)		(35)		(36)		(36)
Weak Ties[a]	15	33	15	33	10	22	3	7	2	4	45	100
		(29)		(29)		(28)		(27)		(25)		(28)
Marketing												
Strong Ties[a]	27	26	25	24	29	28	14	13	9	9	104	100
		(33)		(31)		(32)		(35)		(41)		(33)
Weak Ties[a]	17	32	19	36	13	25	3	6	1	2	53	100
		(33)		(37)		(36)		(27)		(13)		(34)

Notes: a number of firms with corresponding network activities
b percentage figure as row %, percentage figure as column % of respective innovation stage in parentheses

Table A.9 Network activities of manufacturing firms with manufacturing suppliers (1996)

	Metro-politan Scale		National Scale		European Union		Central / Eastern Europe		Mondial Scale		Total	
	no.	%[b]	no.	%[b]	no.	%[b]	no.	%[b]	no.	%[b]	no.	%[b]
Pre-Competitive Stage												
Information Exchange												
Strong Ties[a]	21	27 (39)	24	31 (42)	23	30 (40	5	6 (38)	4	5 (50)	77	100 (41)
Weak Ties[a]	5	33 (21)	3	20 (13)	6	40 (21	1	7 (20)	0	0 (0)	15	100 (17)
Identification of New Ideas												
Strong Ties[a]	16	28 (30)	16	28 (28)	19	33 (33)	4	7 (31)	2	4 (25)	57	100 (30)
Weak Ties[a]	10	29 (42)	11	31 (46)	10	29 (34)	2	6 (40)	2	6 (50)	35	100 (41)
Research & Development												
Strong Ties[a]	17	30 (31)	17	30 (30)	16	29 (28)	4	7 (31)	2	4 (25)	56	100 (29)
Weak Ties[a]	9	25 (38)	10	28 (42)	13	36 (45)	2	6 (40)	2	6 (50)	36	100 (42)
Competitive Stage												
Prototype Development												
Strong Ties[a]	14	25 (44)	17	30 (49)	19	34 (49)	3	5 (50)	3	5 (43)	56	100 (47)
Weak Ties[a]	12	33 (26)	10	28 (22)	10	28 (21)	3	8 (25)	1	3 (20)	36	100 (23)
Pilot Projects												
Strong Ties[a]	11	28 (34)	11	28 (31)	13	33 (33)	2	5 (33)	2	5 (29)	39	100 (33)
Weak Ties[a]	15	28 (33)	16	30 (35)	16	30 (33)	4	8 (33)	2	4 (40)	53	100 (34)
Market Introduction												
Strong Ties[a]	7	29 (22)	7	29 (20)	7	29 (18)	1	4 (17)	2	8 (29)	24	100 (20)
Weak Ties[a]	19	28 (41)	20	29 (43)	22	32 (46)	5	7 (42)	2	3 (40)	68	100 (43)

Notes: a number of firms with corresponding network activities
b percentage figure as row %, percentage figure as column % of respective innovation stage in parentheses

Table A.10 Network activities of manufacturing firms with producer service providers (1996)

	Metro-politan Scale		National Scale		European Union		Central / Eastern Europe		Mondial Scale		Total	
	no.	%[b]	no.	%[b]	no.	%[b]	no.	%[b]	no.	%[b]	no.	%[b]
Pre-Competitive Stage												
Information Exchange												
Strong Ties[a]	38	49 (36)	22	29 (34)	13	17 (26)	3	4 (33)	1	1 (25)	77	100 (33)
Weak Ties[a]	25	47 (30)	15	28 (32)	11	21 (50)	0	0 (0)	2	4 (40)	53	100 (34)
Identification of New Ideas												
Strong Ties[a]	34	45 (32)	20	27 (31)	17	23 (34)	3	4 (33)	1	1 (25)	75	100 (32)
Weak Ties[a]	29	53 (35)	17	31 (36)	7	13 (32)	0	0 (0)	2	4 (40)	55	100 (35)
Research & Development												
Strong Ties[a]	34	42 (32)	22	27 (34)	20	25 (40)	3	4 (33)	2	2 (50)	81	100 (35)
Weak Ties[a]	29	59 (35)	15	31 (32)	4	8 (18)	0	0 (0)	1	2 (20)	49	100 (31)
Competitive Stage												
Prototype Development												
Strong Ties[a]	18	40 (37)	12	27 (31)	12	27 (43)	2	4 (40)	1	2 (33)	45	100 (36)
Weak Ties[a]	45	53 (32)	25	29 (35)	12	14 (27)	1	1 (25)	2	2 (33)	85	100 (32)
Pilot Projects												
Strong Ties[a]	12	35 (24)	13	38 (33)	7	21 (25)	1	3 (20)	1	3 (33)	34	100 (27)
Weak Ties[a]	51	53 (36)	24	25 (33)	17	18 (39)	2	2 (50)	2	2 (33)	96	100 (36)
Market Introduction												
Strong Ties[a]	19	42 (39)	14	31 (36)	9	20 (32)	2	4 (40)	1	2 (33)	45	100 (36)
Weak Ties[a]	44	52 (31)	23	27 (32)	15	18 (34)	1	1 (25)	2	2 (33)	85	100 (32)

Notes: a number of firms with corresponding network activities
b percentage figure as row %, percentage figure as column % of respective innovation stage in parentheses

Table A.11 Network activities of manufacturing firms with research institutions (1996)

	Metro-politan Scale		National Scale		European Union		Central/ Eastern Europe		Mondial Scale		Total	
	no.	%[b]	no.	%[b]	no.	%[b]	no.	%[b]	no.	%[b]	no.	%[b]
Pre-Competitive Stage												
Information Exchange												
Strong Ties[a]	10	40 (30)	5	20 (24)	8	32 (32)	1	4 (100)	1	4 (33)	25	100 (30)
Weak Ties[a]	9	41 (38)	5	23 (56)	5	23 (36)	1	5 (20)	2	9 (33)	22	100 (38)
Identification of New Ideas												
Strong Ties[a]	11	42 (33)	8	31 (38)	7	27 (28)		0 (0)		0 (0)	26	100 (31)
Weak Ties[a]	8	38 (33)	2	10 (22)	6	29 (43)	2	10 (40)	3	14 (50)	21	100 (36)
Research & Development												
Strong Ties[a]	12	38 (36)	8	25 (38)	10	31 (40)		0 (0)	2	6 (67)	32	100 (39)
Weak Ties[a]	7	47 (29)	2	13 (22)	3	20 (21)	2	13 (40)	1	7 (17)	15	100 (26)
Competitive Stage												
Prototype Development												
Strong Ties[a]	10	40 (67)	5	20 (63)	7	28 (54)	1	4 (33)	2	8 (50)	25	100 (58)
Weak Ties[a]	9	41 (21)	5	23 (23)	6	27 (23)	1	5 (33)	1	5 (20)	22	100 (22)
Pilot Projects												
Strong Ties[a]	4	27 (27)	3	20 (38)	5	33 (38)	2	13 (67)	1	7 (25)	15	100 (35)
Weak Ties[a]	15	47 (36)	7	22 (32)	8	25 (31)		0 (0)	2	6 (40)	32	100 (33)
Market Introduction												
Strong Ties[a]	1	33 (7)		0 (0)	1	33 (8)		0 (0)	1	33 (25)	3	100 (7)
Weak Ties[a]	18	41 (43)	10	23 (45)	12	27 (46)	2	5 (67)	2	5 (40)	44	100 (45)

Notes: a number of firms with corresponding network activities
b percentage figure as row %, percentage figure as column % of respective innovation stage in parentheses

Table A.12 Network activities of manufacturing firms with competitors in form of producer networks (1996)

	Metro-politan Scale		National Scale		European Union		Mondial Scale		Total	
	no.	%[b]	no.	%[b]	no.	%[b]	no.	%[b]	no.	%[b]
Pre-Competitive Stage										
Information Exchange										
Strong Ties[a]	14	40	12	34	7	20	2	6	35	100
		(41)		(39)		(30)		(40)		(38)
Weak Ties[a]	4	29	4	29	5	36	1	7	14	100
		(20)		(24)		(38)		(25)		(26)
Identification of New Ideas										
Strong Ties[a]	13	38	12	35	8	24	1	3	34	100
		(38)		(39)		(35)		(20)		(37)
Weak Ties[a]	5	33	4	27	4	27	2	13	15	100
		(25)		(24)		(31)		(50)		(28)
Research & Development										
Strong Ties[a]	7	29	7	29	8	33	2	8	24	100
		(21)		(23)		(35)		(40)		(26)
Weak Ties[a]	11	44	9	36	4	16	1	4	25	100
		(55)		(53)		(31)		(25)		(46)
Competitive Stage										
Prototype Development										
Strong Ties[a]	5	24	6	29	8	38	2	10	21	100
		(33)		(35)		(36)		(33)		(35)
Weak Ties[a]	13	46	10	36	4	14	1	4	28	100
		(33)		(32)		(29)		(33)		(32)
Pilot Projects										
Strong Ties[a]	4	29	3	21	5	36	2	14	14	100
		(27)		(18)		(23)		(33)		(23)
Weak Ties[a]	14	40	13	37	7	20	1	3	35	100
		(36)		(42)		(50)		(33)		(40)
Market Introduction										
Strong Ties[a]	6	24	8	32	9	36	2	8	25	100
		(40)		(47)		(41)		(33)		(42)
Weak Ties[a]	12	50	8	33	3	13	1	4	24	100
		(31)		(26)		(21)		(33)		(28)

Notes: a number of firms with corresponding network activities
b percentage figure as row %, percentage figure as column % of respective innovation stage in parentheses

Table A.13 Selected characteristics of sample producer service firms (1996)

	Sample Firms											
	All		Innov. Firm[a]		Innov. Firm[b]		Non Innov. Firm[c]		Technical Orientated[d]		Business Orientated[e]	
	no.	%	no.	%	no.	%	no.	%	no.	%	no.	%
Corporate Status												
Single Establishment	135	70.7	106	67.9	38	67.9	29	82.9	96	76.2	34	56.7
Multi-Establishment	56	29.3	50	32.1	18	32.1	6	17.1	30	23.8	26	43.3
Main Plant	*33*	*17.3*	*29*	*18.6*	*10*	*17.9*	*4*	*11.4*	*18*	*14.3*	*15*	*25.0*
Branch Plant	*23*	*12.0*	*21*	*13.5*	*8*	*14.2*	*2*	*5.8*	*12*	*9.6*	*11*	*18.4*
Total	191	100	156	100	56	100	35	100	126	100	60	100
Employment Size												
≤ 19	133	71.9	101	66.9	36	66.7	32	94.1	87	71.9	41	69.5
20 – 49	20	16.2	28	18.5	15	27.8	2	5.9	22	18.2	8	13.6
50 – 249	15	8.1	15	9.9	2	3.7	0	0.0	9	7.4	6	10.2
≥ 250	7	3.8	7	4.6	1	1.9	0	0.0	3	2.5	4	6.8
Total	185	100	151	100	54	100	34	100	121	100	59	100
R&D Expenditure[f]												
None	37	21.6	2	1.5	0	0.0	35	100	26	22.2	10	20.0
0.1 – 3.4	106	62.0	106	77.9	33	61.1	0	0.0	69	59.0	35	70.0
3.5 – 7.9	18	10.5	18	13.2	14	25.9	0	0.0	16	13.7	2	4.0
≥ 8	10	5.8	10	7.4	7	13.0	0	0.0	6	5.1	3	6.0
Total	171	100	136	100	54	100	35	100	117	100	50	100

Notes: a introducing new services and/or organizational innovations
b spending more than 20% of turnover in R&D activities
c spending less than 20% of turnover in R&D activities
d belonging to the sectors of computer software and technical consultancy
e belonging to the sectors of business consultancy and market research/advertising
f annual average between 1994 - 1996

Table A.14 R&D expenses of producer service firms (1996) disaggregated by size classes

		Firm Size by Employees									
		≤ 19		20 – 49		50 – 249		≥ 250		Total	
		no.	%[b]	no.	%[b]	no.	%[b]	no.	%[b]	no.	%[b]
***R&D Expenses* (in % of turnover)[a]**											
None		33	27.3	2	6.9	0	0.0	1	25.0	36	21.7
	c	91.7		5.6		0.0		2.8		100	
0.1 – 3.4		68	56.2	22	75.9	10	83.3	2	50.0	102	61.4
	c	66.7		21.6		9.8		2.0		100	
3.5 – 7.9		12	9.9	3	10.3	2	16.7	1	25.0	18	10.8
	c	66.7		16.7		11.1		5.6		100	
≥ 8		8	6.6	2	6.9	0	0.0	0	0.0	10	6.0
	c	80.0		20.0		0.0		0.0		100	
Total		121	100	29	100	12	100	4	100	166	100
	c	72.9		17.5		7.2		2.4		100	

Notes: a average 1994 to 1996
b percentage figure as column %
c percentage figure as row %

Table A.15 Innovativeness of producer service firms (1996) by age

	Year of Founding / Establishment									
	Before 1970		1970-1979		1980-1989		Since 1990		All	
	no.	%[a]	no.	%[a]	no.	%[a]	no.	%[a]	no.	%[a]
Firms										
Non-Innovative	4	11.8	7	20.6	13	38.2	10	29.4	34	100
Innovative[b]	27	17.5	33	21.4	55	35.7	39	25.3	154	100
Total	31	16.5	40	21.3	68	36.2	49	26.1	188	100

Notes: a percentage figure as row %
b firms with product innovations in 1994-1996

Table A.16 Firms with service innovations (1994 - 1996)

	Single Establishment		Main Plant		Branch Plant		Total	
	no.	%[a]	no.	%[a]	no.	%[a]	no.	%[a]
None	29	82.9	4	11.4	2	5.8	35	100
New Services	48	60.0	20	25.0	12	15.0	80	100
Substantially Improved Services	45	61.6	18	24.7	10	13.7	73	100
New or Substantially Improved Methods of Services Provision	84	69.4	22	18.2	15	12.4	121	100

Note: a percentage figure as row %

Table A.17 Types of innovation support for manufacturing firms (1996) by producer service firms

	Computer Software		Technical Consultancy		Business Consultancy		Market Research and Advertising		Total	
In %	yes	no	yes	no	yes	no	yes	no	yes	no
Product Innovation	75.0	25.0	67.4	32.6	55.0	45.0	77.3	22.7	68.3	31.7
Process Innovation	100.0	0.0	56.5	43.5	75.0	25.0	45.5	54.5	64.4	35.6
Organisat. Innovations	81.3	18.8	50.0	50.0	85.0	15.0	72.7	27.3	65.3	33.7
New Sales Markets with Exist. Products	43.8	56.3	26.1	73.9	50.0	50.0	90.9	9.1	47.1	52.9

Table A.18 Important factors for successful collaboration with manufacturing clients (1996)

	Computer Software		Technical Consultancy		Business Consultancy		Market Research and Advertising		Total	
	no.	%[a]	no.	%[a]	no.	%[a]	no.	%[a]	no.	%[a]
Factors										
Spatial Proximity	7	15.9	13	11.5	3	6.8	6	11.3	29	11.4
b		24.1		44.8		10.3		20.7		100
Frequent Personal Contacts	15	34.1	39	34.5	17	38.6	17	32.1	88	34.6
b		17.0		44.3		19.3		19.3		100
Existence of Similar Qualifications	10	22.7	26	23.0	9	20.5	11	20.8	56	22.0
b		17.9		46.4		16.1		19.6		100
Good Knowledge of Client's Industry	12	27.3	35	31.0	15	34.1	19	35.8	81	31.9
b		14.8		43.2		18.5		23.5		100
Total	44	100	113	100	44	100	53	100	254	100
b		17.3		44.5		17.3		20.9		100

Notes: **a percentage figure as column %**
b percentage figure as row %

Table A.19 Establishments of contacts (1996)

	Computer Software		Technical Consultancy		Business Consultancy		Market Research and Advertising		Total	
	no.	%[a]	no.	%[a]	no.	%[a]	no.	%[a]	no.	%[a]
Contacts through										
Advertising/ Marketing	13	25.0	18	9.8	8	15.7	25	29.8	64	17.3
Conference/ Meetings	6	11.5	25	13.6	7	13.7	13	15.5	51	13.7
Fairs	4	7.7	12	6.5	1	1.9	5	5.9	22	5.9
References/ Recommendations	17	3.3	76	41.3	23	45.1	25	29.8	141	38.0
Contact Agencies	1	1.9	8	4.3	1	1.9	1	1.2	11	2.9
Existing Contacts w. Former Colleagues	7	13.5	35	19.0	8	15.7	10	11.9	60	16.1
Existing Contacts w. Former Employees	4	7.7	10	5.4	3	5.9	5	5.9	22	5.9
Total	52	100	184	100	51	100	84	100	371	99.8

Note: a percentage in % of responses

Table A.20 Relevance of personal contacts (1996)

	Metropolitan Scale		National Scale		International Scale		Not Important	
	no.	%[a]	no.	%[a]	no.	%[a]	no.	%[a]
Contacts from the Time of								
Studies	88	38.4	49	33.8	19	20.9	44	14.6
Doct./Post Doc. at Univ.	23	10.0	12	8.3	8	8.8	117	38.8
Professional Activities in								
Research Establishments	24	10.5	14	9.7	18	19.8	112	37.2
Business	94	41.0	70	48.3	46	50.5	28	9.3
Total	229	100	145	100	91	100	301	100

Note: a percentage figure as column %

Table A.21 Selected characteristics of sample research establishments (1996)

	All		Research Intensive[a]		Non-Research Intensive[b]	
	no.	%	no.	%	no.	%
Status						
University	132	45.5	66	35.3	61	70.9
Public Non-Univ. Organisations[c]	158	54.5	121	64.7	25	29.1
Total	290	100.0	187	100.0	86	100.0
Employment Size						
≤ 19	234	82.4	150	81.5	69	82.1
20 – 49	36	12.7	25	13.6	11	13.1
≥ 50	14	4.8	9	4.9	4	4.8
Total	284	100.0	184	100.0	84	100.0

Notes: a more than 50% of the total time budget of all scientific staff devoted to basic or applied research
b less than 50% of the total time budget of all scientific staff devoted to basic or applied research
c including business-related research centres

Table A.22 Reasons for collaboration with research establishments (1996)

	Research Establishments in Science Fields											
	Architect. Construct., Surveying		Biology, Chemistry Medicine		Mathem., Inform., Physics		Electrotechn. Mechanical Engineering		Economics, Social and Geosciences		Total	
	no.	%[a]	no.	%[a]	no.	%[a]	no.	%[a]	no.	%[a]	no.	%[a]
Insufficient Own Equipment	11	58	36	49	8	28	14	56	43	58	112	51
Insufficient Own Personnel Capacity	10	53	40	55	13	45	13	52	45	61	121	55
Financial Sponsorship only Avail. for Collab.	15	79	25	34	16	55	11	44	33	45	100	45
Ideas for Research Work/Thematic Additions	15	79	59	81	25	86	20	80	59	80	178	81
Raising Own Profile	3	16	24	33	9	31	5	20	30	40	71	32
n	19		73		29		25		74		220	

Note: a percentage figure as % of cases

Table A.23 Branch structure of business clients (1996)

	Research Establishments in Science Fields											
	Architect.; Construct., Surveying		Biology, Chemistry Medicine		Mathem., Inform., Physics		Electrotechn. Mechanical Engineering		Economics, Social and Geosciences		Total	
	no.	%[a]	no.	%[a]	no.	%[a]	no.	%[a]	no.	%[a]	no.	%[a]
Energy and Mining	4	25	2	4	4	18	7	24	10	26	27	18
Basic Metals and Metal Products	2	13	7	15	2	9	12	41	1	3	24	16
Chemicals	2	13	29	60	2	9	1	3	3	8	37	24
Electrical & Optical Equipment	1	6	13	27	8	36	18	62	5	13	45	29
Computers	3	19	8	17	11	50	7	24	6	15	35	23
Plastics & Rubber	2	13	6	13	1	5	0	0	2	5	11	7
Machinery & Transport	5	31	3	6	3	14	13	45	4	10	28	18
Wood, Paper & Printing	1	6	1	2	2	9	0	0	4	10	8	5
Textiles & Clothing	0	0	1	2	0	0	0	0	0	0	1	1
Food Industry	0	0	5	10	1	4	0	0	4	10	10	7
Service Industry	5	31	4	8	4	18	4	14	28	72	45	29
n	16		48		22		29		39		154	

Note: a percentage figure as % of cases

Table A.24 Primary reasons for collaboration with businesses (1996)

	Research Establishments in Science Fields											
	Architect.; Construct., Surveying		Biology, Chemistry Medicine		Mathem., Inform., Physics		Electrotechn., Mechanical Engineering		Economics, Social and Geosciences		Total	
	no.	%[a]	no.	%[a]	no.	%[a]	no.	%[a]	no.	%[a]	no.	%[a]
Enables Costly Research Projects	11	58	37	75	7	32	18	62	21	52	94	59
Receipt of Funds Tied to Cooperation	11	58	25	51	12	45	20	69	20	50	88	55
Reduces Dependency on Public Contracts	8	42	27	55	10	45	8	28	19	74	72	45
Practice-Oriented Impulses for the Research Project	19	100	36	73	19	86	25	86	39	97	138	87
Creation of Jobs for Scientific Qualification	9	47	20	41	10	45	10	34	9	22	58	36
Use of Businesses' Capacities	10	53	22	45	2	9	12	41	16	40	62	39
n	19		49		22		29		40		159	

Note: a percentage figure as % of cases

Table A.25 Primary channels for establishing contacts (1996)

	Research Establishment in Science Fields											
	Architect.; Construct., Surveying		Biology, Chemistry Medicine		Mathem., Inform., Physics		Electrotech., Mechanical Engineering		Economics, Social and Geosciences		Total	
	no.	%[a]	no.	%[a]	no.	%[a]	no.	%[a]	no.	%[a]	no.	%[a]
Through												
Congresses/Fairs	4	22.2	18	37.5	6	27.3	6	20.7	8	20.0	42	26.8
Specialist Magazines	6	33.3	10	20.8	3	13.6	4	13.8	9	22.5	32	20.4
Data Banks	0	0.0	1	2.1	2	9.1	0	0.0	3	7.5	6	3.8
Associations	5	27.8	1	2.1	0	0.0	6	20.7	10	25.0	22	14.0
Transfer Establ./ Contact Agencies	2	11.1	0	0.0	1	4.5	4	13.8	4	10.0	11	7.0
Approach by Businesses	11	61.1	33	68.8	15	68.2	21	72.4	24	60.0	104	66.2
Personal Contact of Staff	17	94.4	42	87.5	19	86.4	23	79.3	35	87.5	136	86.6
n	18		48		22		29		40		157	

Note: a percentage figure as % of cases

Table A.26 Problems during collaboration with businesses (1996)

	Manufacturing Firms		Producer Service Firms		Total	
	no.	%[a]	no.	%[a]	no.	%[a]
Problems with Project Leadership	16	42.9	3	20.8	19	22.6
Budgeted Cost Overrun	13	14.3	1	16.9	14	16.7
Coordination is Difficult	43	71.4	5	55.8	38	57.1
Different Capabilities	37	85.7	6	48.1	43	51.2
Lack of Schedule Effectiveness	23	57.1	4	29.9	27	32.1
n	77		7		84	

Note: a percentage figure as % of cases

Table A.27 Barriers before collaborating with businesses/enterprises (1996)

	Cooperation with					
	Manufacturing Firms		Producer Service Firm		Total	
	no.	%[a]	no.	%[a]	no.	%[a]
Lacking of Actual Contact Partner	33	32.0	3	50.0	36	33.0
Financial Budget of Businesses Insufficient	74	71.8	5	83.3	79	72.5
Fear of Knowledge Drain to:						
the Businesses	45	43.7	1	16.7	46	42.2
the Research Establishments	18	17.5	0	0.0	18	16.5
n	103		6		109	

Note: a percentage figure as % of cases

Table A.28 Relevance of personal contacts (1996)

	Metropolitan Scale		National Scale		International Scale	
	no.	%[a]	no.	%[a]	no.	%[a]
Contacts from the Time of						
Studies	140	53.4	89	34.0	50	19.1
Ph.D./Post Doc Studies	79	30.2	58	22.1	70	26.7
Professional Activities in						
Research Institutions	175	66.8	145	55.3	151	57.6
Business	49	18.7	48	18.3	56	19.1

Note: a percentage figure as % of cases

APPENDIX B

The Barcelona Metropolitan Innovation System

Manufacturing Sector:	Tables B.1 – B.12
Producer Service Sector:	Tables B.13– B.20
Science & Research Sector:	Tables B.21– B.28

Table B.1 Selected characteristics of manufacturing sample firms (1996)

	Sample Firms with					
	Regular R&D[b]		No R&D[c]		Total	
	no.	%	no.	%	no.	%
Corporate Status						
Single Establishment	164	71.3	118	87.4	282	77.3
Multi-Establishment	66	28.7	17	12.6	83	22.7
Main Plant	*39*	*17.0*	*10*	*7.4*	*49*	*13.4*
Branch Plant	*27*	*11.7*	*7*	*5.2*	*34*	*9.3*
Total	230	100	135	100	365	100
Employment Size						
≤ 49	116	50.4	105	77.2	221	60.4
50 – 99	55	23.9	21	15.4	76	20.8
100 – 499	42	18.3	10	7.4	52	14.2
≥ 500	17	7.4	0	0.0	17	4.6
Total	230	100	136	100	366	100
R&D Expenditure[a]						
None	5	2.4	50	42.0	55	16.8
0.1 – 3.4	92	44.2	52	43.7	144	44.0
3.5 – 7.9	51	24.5	11	9.2	62	19.0
≥ 8	60	28.8	6	5.0	66	20.2
Total	208	100	119	100	327	100

Notes: a annual average of 1994-1996
b regular R&D firms are permanently engaged in research or development process
c no R&D firms are only occasionally or never engaged in research or development process

Table B.2 R&D of manufacturing firms (1996) disaggregated by size classes

	Firm Size by Employees									
	≤ 49		50 – 99		100 – 499		≥ 500		Total	
	no.	%[b]	no.	%[b]	no.	%[b]	no.	%[b]	no.	%[b]
R&D Expenses **(in % of turnover)**										
No R&D	49	25.1	4	6.0	4	8.2	0	0.0	57	17.4
0.1 - 3.4	71	36.4	33	49.3	30	61.2	9	52.9	143	43.6
3.5 - 7.9	40	20.5	11	16.4	9	18.4	4	23.5	64	19.5
≥ 8	35	17.9	19	28.4	6	12.2	4	23.5	64	19.5
Total	195	100	67	100	49	100	17	100	328	100

Notes: a average 1994-1996
b percentage figure as column %

Table B.3 Innovativeness of manufacturing firms (1996) by age

	Year of Founding /Establishment									
	Before 1969		1970-1979		1980-1989		Since 1990		All	
	no.	%[a]	no.	%[a]	no.	%[a]	no.	%[a]	no.	%[a]
Firms										
Non-Innovative	19	26.4	17	23.6	17	23.6	19	26.4	72	100
Innovative[b]	78	27.5	61	21.5	71	25.0	74	26.1	284	100
Total	97	27.2	78	21.9	88	24.7	93	26.1	356	100

Notes: a percentage figure as row %
b firms with product innovations in 1994-1996

Table B.4 Manufacturing firms with product innovations (1994 - 1996)

	Corporate Status							
	Single Establishment		Main Plant		Branch Plant		Total	
	no.	%[a]	no.	%[a]	no.	%[a]	no.	%[a]
None	66	86.8	6	7.9	4	5.3	76	100
Further Development of Products[b]	92	72.4	18	14.2	17	13.4	127	100
New Products[c]	70	77.8	14	15.6	6	6.7	90	100
Total	228	77.8	38	13.0	27	9.2	293	100

Notes: a percentage figure as row %
b more than 50% of the products were substantially developed further during the years 1994 to 1996
c more than 50% of the were newly introduced in the product programme 1994 to 1996

Table B.5 Sources of external information for product and process innovation (1996)

	Product Innovation Activities		Process Innovation Activities	
	no.	%[a]	no.	%[b]
Sources				
Customers	240	84.2	72	28.2
Suppliers	124	43.5	159	62.4
Competitors	155	54.4	66	25.9
Research Institutions	57	20.0	54	21.2
Producer Services	77	27.0	108	42.4
Fairs/Exhibitions	197	69.1	150	58.8
Specialised Literature	110	38.6	101	39.6
Media	53	18.6	39	15.3
Internet	26	9.1	12	4.7
n	285		255	

Notes: a percentage of all firms with product innovation activities
b percentage of all firms with process innovation activities

Table B.6 Motives for exercising network activities (1996)

	Network Activities with					
	Research Institutions		Producer Service Providers		Other Manufacturing Firms[b]	
	no.	%[a]	no.	%[a]	no.	%[a]
Motives						
Risks/Cost Reduction	15	17.2	55	40.7	97	66.4
New Technological Opportunities	59	67.8	60	44.5	60	41.4
Know-How-Takeover	42	48.2	39	28.9	43	29.5
Financial Resources	3	3.4	56	41.4	9	6.2
Funding Requirements	15	17.2	61	38.9	3	2.1
n	87		1,587		146	

Notes: a percentage of manufacturing firms cooperating with the corresponding network partner
b relation of manufacturing firms with manufacturing suppliers, customers or competitors

Table B.7 Problems with exercising network activities (1996)

	Network Activities with					
	Research Institutions		Producer Service Providers		Other Manufacturing Firms[b]	
	no.	%[a]	no.	%[a]	no.	%[a]
Problems						
Problem with Project Management	10	19.2	27	24.5	21	17.1
Budgeted Cost Overrun	13	25.0	54	49.1	37	30.1
Unintended Knowledge Drain	5	9.6	7	6.4	10	8.1
Coordination Difficult	24	46.2	34	30.9	46	37.4
Different Capability	17	32.6	26	23.6	31	25.2
Confidential Relation/ Secrecy	2	3.8	5	4.5	12	9.8
Loss of Independence	3	5.8	6	5.5	16	13.0
Lack of Schedule Effectiveness	18	34.6	42	38.2	91	74.0
n	52		110		123	

Notes: a percentage of manufacturing firms cooperating with the corresponding network partner
b relation of manufacturing firms with manufacturing suppliers, customers or competitors

Table B.8 Network activities of manufacturing firms with customers (1996)

	Metro-politan Region		Catalonia		National Scale		European Union		Mondial Scale		Total	
	no.	%[b]	no.	%[b]	no.	%[b]	no.	%[b]	no.	%[b]	no.	%[b]
Pre-competitive Stage												
Information Exchange												
Strong Ties[a]	128	29	79	18	99	23	69	16	65	15	440	100
		(42)		(39)		(39)		(40)		(42)		(40)
Weak Ties[a]	30	34	19	22	20	23	11	13	8	9	88	100
		(21)		(23)		(23)		(17)		(14)		(20)
Identification of New Ideas												
Strong Ties[a]	104	27	71	19	90	24	58	15	57	15	380	100
		(34)		(35)		(36)		(34)		(37)		(35)
Weak Ties[a]	50	35	27	19	29	20	22	15	16	11	144	100
		(35)		(33)		(33)		(34)		(29)		(33)
R&D												
Strong Ties[a]	74	28	54	20	63	24	44	16	33	12	268	100
		(24)		(26)		(25)		(26)		(21)		(25)
Weak Ties[a]	62	31	36	18	39	20	31	16	32	16	200	100
		(44)		(44)		(44)		(48)		(57)		(46)
Competitive Stage												
Prototype Development												
Strong Ties[a]	96	29	66	20	76	23	48	15	45	14	331	100
		(36)		(36)		(34)		(35)		(34)		(35)
Weak Ties[a]	52	30	33	19	37	21	27	15	26	15	175	100
		(30)		(32)		(32)		(30)		(31)		(31)
Pilot Projects												
Strong Ties[a]	90	28	65	20	76	24	49	15	38	12	318	100
		(34)		(35)		(34)		(36)		(29)		(34)
Weak Ties[a]	53	30	28	16	34	20	28	16	31	18	174	100
		(31)		(27)		(30)		(31)		(37)		(31)
Marketing												
Strong Ties[a]	80	27	54	18	72	24	41	14	48	16	295	100
		(30)		(29)		(32)		(30)		(37)		(31)
Weak Ties[a]	68	32	41	19	44	20	36	17	26	12	215	100
		(39)		(40)		(38)		(40)		(31)		(38)

Notes: a number of firms with corresponding network activities
b percentage figure as row %, percentage figure as column % of respective innovation stage in parentheses

Table B.9 Network activities of manufacturing firms with manufacturing suppliers (1996)

	Metro-politan Region		Catalonia		National Scale		European Union		Mondial Scale		Total	
	no.	%[b]	no.	%[b]	no.	%[b]	no.	%[b]	no.	%[b]	no.	%[b]
Pre-competitive Stage												
Information Exchange												
Strong Ties[a]	99	37 (39)	52	19 (39)	46	17 (39)	46	17 (38)	25	9 (40)	268	100 (39)
Weak Ties[a]	33	39 (28)	16	19 (27)	13	15 (26)	19	23 (29)	3	4 (18)	84	100 (27)
Identification of New Ideas												
Strong Ties[a]	89	36 (35)	47	19 (36)	42	17 (35)	42	17 (35)	25	10 (40)	245	100 (36)
Weak Ties[a]	32	37 (27)	17	20 (28)	14	16 (28)	21	24 (32)	3	3 (18)	87	100 (28)
R&D												
Strong Ties[a]	63	36 (25)	33	19 (25)	31	18 (26)	33	19 (27)	13	8 (21)	173	100 (25)
Weak Ties[a]	54	38 (45)	27	19 (45)	23	16 (46)	26	18 (39)	11	8 (65)	141	100 (45)
Competitive Stage												
Prototype Development												
Strong Ties[a]	75	37 (51)	41	20 (47)	33	16 (43)	39	19 (48)	15	7 (42)	203	100 (47)
Weak Ties[a]	48	39 (25)	21	17 (24)	22	18 (27)	23	19 (24)	10	8 (28)	124	100 (25)
Pilot Projects												
Strong Ties[a]	49	35 (33)	27	19 (31)	25	18 (32)	27	19 (33)	11	8 (31)	139	100 (32)
Weak Ties[a]	62	38 (32)	29	18 (33)	27	17 (33)	32	20 (34)	13	8 (36)	163	100 (33)
Marketing												
Strong Ties[a]	24	27 (16)	19	22 (22)	19	22 (25)	16	18 (20)	10	11 (28)	88	100 (20)
Weak Ties[a]	85	41 (44)	38	18 (43)	32	15 (40)	40	19 (42)	13	6 (36)	208	100 (42)

Notes: a number of firms with corresponding network activities
b percentage figure as row %, percentage figure as column % of respective innovation stage in parentheses

Table B.10 Network activities of manufacturing firms with producer service providers (1996)

	Metro-politan Region		Catalonia		National Scale		European Union		Mondial Scale		Total	
	no.	%[b]	no.	%[b]	no.	%[b]	no.	%[b]	no.	%[b]	no.	%[b]
Pre-competitive Stage												
Information Exchange												
Strong Ties[a]	100	48 (39)	43	20 (38)	40	19 (37)	21	10 (32)	6	3 (32)	210	100 (37)
Weak Ties[a]	33	49 (25)	13	19 (25)	10	15 (26)	9	13 (30)	2	3 (33)	67	100 (26)
Identification of New Ideas												
Strong Ties[a]	80	46 (31)	33	19 (29)	33	19 (31)	20	12 (31)	7	4 (37)	173	100 (31)
Weak Ties[a]	48	49 (36)	21	21 (40)	15	15 (38)	12	12 (40)	2	2 (33)	98	100 (37)
R&D												
Strong Ties[a]	78	43 (30)	37	21 (33)	35	19 (32)	24	13 (37)	6	3 (32)	180	100 (32)
Weak Ties[a]	53	55 (40)	19	20 (36)	14	14 (36)	9	9 (30)	2	2 (33)	97	100 (37)
Competitive Stage												
Prototype Development												
Strong Ties[a]	69	42 (43)	38	23 (43)	34	20 (43)	21	13 (48)	4	2 (36)	166	100 (44)
Weak Ties[a]	58	54 (28)	18	17 (24)	14	13 (22)	13	12 (25)	5	5 (36)	108	100 (26)
Pilot Projects												
Strong Ties[a]	53	41 (33)	30	23 (34)	27	21 (34)	15	12 (34)	3	2 (27)	128	100 (34)
Weak Ties[a]	67	51 (33)	24	18 (32)	20	15 (32)	16	12 (31)	5	4 (36)	132	100 (32)
Marketing												
Strong Ties[a]	37	43 (23)	20	23 (23)	18	21 (23)	8	9 (18)	4	5 (36)	87	100 (23)
Weak Ties[a]	80	47 (39)	34	20 (45)	29	17 (46)	22	13 (43)	4	2 (29)	169	100 (41)

Notes: a number of firms with corresponding network activities
b percentage figure as row %, percentage figure as column % of respective innovation stage in parentheses

Table B.11 Network activities of manufacturing firms with research institutions (1996)

	Metro-politan Region		Catalonia		National Scale		European Union		Mondial Scale		Total	
	no.	% [b]	no.	% [b]	no.	% [b]	no.	% [b]	no.	% [b]	no.	% [b]
Pre-competitive Stage												
Information Exchange												
Strong Ties [a]	28	37	9	12	19	25	13	17	7	9	76	100
		(31)		(31)		(40)		(34)		(44)		(34)
Weak Ties [a]	18	64	4	14	3	11	3	11		0	28	100
		(38)		(44)		(23)		(38)		(0)		(33)
Identification of New Ideas												
Strong Ties [a]	27	42	8	12	14	22	11	17	5	8	65	100
		(30)		(28)		(29)		(29)		(31)		(29)
Weak Ties [a]	19	56	4	12	5	15	3	9	3	9	34	100
		(40)		(44)		(38)		(38)		(50)		(40)
R&D												
Strong Ties [a]	36	44	12	15	15	19	14	17	4	5	81	100
		(40)		(41)		(31)		(37)		(25)		(36)
Weak Ties [a]	11	50	1	5	5	23	2	9	3	14	22	100
		(23)		(11)		(38)		(25)		(50)		(26)
Competitive Stage												
Prototype Development												
Strong Ties [a]	30	41	8	11	16	22	14	19	6	8	74	100
		(56)		(62)		(62)		(58)		(60)		(58)
Weak Ties [a]	16	53	4	13	5	17	4	13	1	3	30	100
		(22)		(17)		(17)		(18)		(10)		(19)
Pilot Projects												
Strong Ties [a]	20	45	5	11	9	20	7	16	3	7	44	100
		(37)		(38)		(35)		(29)		(30)		(35)
Weak Ties [a]	23	43	8	15	10	19	8	15	4	8	53	100
		(32)		(35)		(34)		(36)		(40)		(34)
Marketing												
Strong Ties [a]	4	44		0	1	11	3	33	1	11	9	100
		(7)		(0)		(4)		(13)		(10)		(7)
Weak Ties [a]	34	46	11	15	14	19	10	14	5	7	74	100
		(47)		(48)		(48)		(45)		(50)		(47)

Notes: a number of firms with corresponding network activities
b percentage figure as row %, percentage figure as column % of respective innovation stage in parentheses

Table B.12 Network activities of manufacturing firms with competitors in form of producer networks (1996)

	Metro-politan Region		Catalonia		National Scale		European Union		Mondial Scale		Total	
	no.	% [b]	no.	% [b]	no.	% [b]	no.	% [b]	no.	% [b]	no.	% [b]
Pre-competitive Stage												
Information Exchange												
Strong Ties [a]	29	34 (38)	12	14 (31)	15	17 (34)	20	23 (37)	10	12 (38)	86	100 (36)
Weak Ties [a]	6	22 (21)	7	26 (33)	5	19 (29)	5	19 (26)	4	15 (50)	27	100 (29)
Identification of New Ideas												
Strong Ties [a]	24	29 (32)	14	17 (36)	14	17 (32)	19	23 (35)	11	13 (42)	82	100 (34)
Weak Ties [a]	9	32 (32)	6	21 (29)	6	21 (35)	6	21 (32)	1	4 (13)	28	100 (30)
R&D												
Strong Ties [a]	23	32 (30)	13	18 (33)	15	21 (34)	15	21 (28)	5	7 (19)	71	100 (30)
Weak Ties [a]	13	34 (46)	8	21 (38)	6	16 (35)	8	21 (42)	3	8 (38)	38	100 (41)
Competitive Stage												
Prototype Development												
Strong Ties [a]	16	25 (40)	12	19 (46)	14	22 (44)	16	25 (43)	5	8 (38)	63	100 (43)
Weak Ties [a]	15	37 (29)	9	22 (31)	6	15 (27)	7	17 (23)	4	10 (33)	41	100 (28)
Pilot Projects												
Strong Ties [a]	12	24 (30)	9	18 (35)	10	20 (31)	13	27 (35)	5	10 (38)	49	100 (33)
Weak Ties [a]	19	37 (37)	10	19 (34)	8	15 (36)	11	21 (35)	4	8 (33)	52	100 (36)
Marketing												
Strong Ties [a]	12	33 (30)	5	14 (19)	8	22 (25)	8	22 (22)	3	8 (23)	36	100 (24)
Weak Ties [a]	17	33 (33)	10	19 (34)	8	15 (36)	13	25 (42)	4	8 (33)	52	100) (36)

Notes: a number of firms with corresponding network activities
b percentage figure as row %, percentage figure as column % of respective innovation stage in parentheses

Table B.13 Selected characteristics of sample producer service firms (1996)

	Sample Firms											
	All		Innov. Firm[a]		Innov. Firm[b]		Non Innov. Firm[c]		Technical Orientated[d]		Business Orientated[e]	
	no.	%	no.	%	no.	%	no.	%	no.	%	no.	%
Corporate Status												
Single Establishment	68	57.6	63	56.3	19	59.4	5	83.3	37	56.9	31	58.8
Multi-Establishment	50	42.4	59	43.7	13	40.6	1	16.7	28	43.1	22	41.5
Main Plant	*36*	*30.5*	*36*	*32.1*	*10*	*31.3*	*0*	*0.0*	*20*	*30.8*	*16*	*30.2*
Branch Plant	*14*	*11.8*	*13*	*11.6*	*3*	*9.4*	*1*	*16.7*	*8*	*12.3*	*6*	*11.3*
Total	168	100	112	100	31	100	6	100	65	100	53	100
Employment Size												
≤ 19	45	39.1	42	38.2	12	37.5	3	60	27	41.5	18	36.0
20 – 49	48	41.7	46	41.8	15	46.9	2	40	27	41.5	21	42.0
50 – 249	16	13.9	16	14.5	5	15.6	0	0.0	10	15.4	6	12.0
≥ 250	6	5.2	6	5.5	0	0.0	0	0.0	1	1.5	5	10.0
Total	115	100	110	100	32	100	5	100	65	100	50	100
R&D Expenditure[f]												
None	1	1.0	1	1.0	0	0.0	6	100	0	0.0	1	2.4
0.1 – 3.4	66	68.0	66	68.0	15	48.4	0	0.0	37	66.1	29	70.7
3.5 – 7.9	20	20.6	20	20.6	8	25.8	0	0.0	11	19.6	9	22.0
≥ 8	10	10.3	10	10.3	8	25.8	0	0.0	8	14.3	2	4.9
Total	97	100	97	100	31	100	0	100	56	100	41	100

Notes: a introducing new services and/or organisational innovations
b spending more than 20% of turnover in R&D activities
c spending less than 20% of turnover in R&D activities
d belonging to the sectors of computer software and technical consultancy
e belonging to the sectors of business consultancy and market research/advertising
f annual average between 1994 - 1996

Table B.14 R&D expenses of producer service firms (1996) disaggregated by size classes

		Firm Size by Employees									
		≤ 19		20 – 49		50 – 249		≥ 250		Total	
		no.	%[b]	no.	%[b]	no.	%[b]	no.	%[b]	no.	%[b]
***R&D Expenses* (in % of turnover)[a]**											
None		0	0.0	1	2.4	0	0.0	0	0.0	1	1.1
	c	0.0		100		0.0		0.0		100	
0.1 – 3.4		23	60.5	29	70.7	11	84.6	3	100	46	69.5
	c	34.8		43.9		16.7		4.5		100	
3.5 – 7.9		9	23.7	10	24.4	0	0.0	0	0.0	19	20.2
	c	47.4		52.6		0.0		0.0		100	
≥ 8		6	15.8	1	2.4	2	15.4	0	0.0	9	9.5
	c	66.7		11.1		22.2		0.0		100	
Total		38	100	41	100	13	100	3	100	95	100
	c	40.0		43.2		13.7		3.2		100	

Notes: a average 1994 to 1996
b percentage figure as column %
c percentage figure as row %

Table B.15 Innovativeness of producer service firms (1996) by age

	Year of Founding / Establishment									
	Before 1970		1970-1979		1980-1989		Since 1990		All	
	no.	%[a]	no.	%[a]	no.	%[a]	no.	%[a]	no.	%[a]
Firms										
Non-Innovative	0	0.0	0	0.0	1	16.7	5	83.3	6	100
Innovative[b]	10	9.0	7	6.3	35	31.5	59	53.2	111	100
Total	10	8.5	7	6.0	36	30.8	64	54.7	117	100

Notes: a percentage figure as row %
b firms with product innovations in 1994-1996

Table B.16 Firms with service innovations (1994 - 1996)

	Single Establishment		Main Plant		Branch Plant		Total	
	no.	%[a]	no.	%[a]	no.	%[a]	no.	%[a]
None	29	82.9	4	11.4	2	5.8	35	100
New Services	32	55.2	20	34.5	6	10.3	58	100
Substantially Improved Services	31	58.5	15	28.3	7	13.2	53	100
New or Substantially Improved Methods of Services Provision	34	49.3	24	34.8	11	15.9	69	100

Note: a percentage figure as row %

Table B.17 Types of innovation support for manufacturing firms (1996) by producer service firms

	Computer Software		Technical Consultancy		Business Consultancy		Market Research and Advertising		Total	
In %	yes	no	yes	no	yes	no	yes	no	yes	no
Product Innovation	33.3	66.7	51.6	48.4	36.4	63.6	62.5	37.6	45.6	54.4
Process Innovation	50.5	9.5	77.4	22.6	68.2	31.8	50.0	50.0	73.3	26.7
Organisat. Innovations	33.3	66.7	77.4	22.6	77.3	22.7	56.3	43.8	63.3	36.7
New Sales Markets with Exist. Products	33.3	66.7	48.4	51.6	40.9	59.1	52.5	37.5	45.6	54.4

Table B.18 Important factors for successful collaboration with manufacturing clients (1996)

		Computer Software		Technical Consultancy		Business Consultancy		Market Research and Advertising		Total	
		no.	%[a]	no.	%[a]	no.	%[a]	no.	%[a]	no.	%[a]
Factors											
Spatial Proximity		7	12.3	10	13.0	7	14.0	6	18.2	30	13.8
	b		23.3		33.3		23.3		20.0		100
Frequent Personal Contacts		18	31.6	22	28.6	17	34.0	14	42.4	71	32.7
	b		25.3		31.0		23.9		19.7		100
Existence of Similar Qualifications		15	26.3	21	27.3	12	24.0	4	12.1	52	23.9
	b		28.8		40.4		23.1		7.7		100
Good Knowledge of Client's Industry		17	29.8	24	31.2	14	28.0	9	27.3	64	29.5
	b		26.6		37.5		21.8		14.1		100
Total		57	100	77	100	50	100	33	100	217	100
	b		26.3		35.5		23.0		15.2		100

Notes: a percentage figure as column %
b percentage figure as row %

Table B.19 Establishments of contacts (1996)

	Computer Software		Technical Consultancy		Business Consultancy		Market Research and Advertising		Total	
	no.	%[a]	no.	%[a]	no.	%[a]	no.	%[a]	no.	%[a]
Contacts through										
Advertising/ Marketing	30	26.3	8	13.3	10	13.2	12	26.7	60	20.3
Conference/ Meetings	14	12.3	11	18.3	12	15.8	5	11.1	42	14.2
Fairs	14	12.3	7	11.7	6	7.9	2	4.4	29	9.8
References/ Recommendations	30	26.3	17	28.3	23	30.3	11	24.4	81	27.5
Contact Agencies	0	0.0	1	1.7	1	1.3	0	0.0	2	0.7
Existing Contacts w. Former Colleagues	14	12.3	8	13.3	15	19.7	8	17.8	45	15.3
Existing Contacts w. Former Employees	12	10.5	8	13.3	9	11.8	7	15.6	36	12.2
Total	114	100	60	100	76	100	45	100	295	100

Note: a percentage in % of responses

Table B.20 Relevance of personal contacts (1996)

	Metropolitan Scale		National Scale		International Scale		Not Important	
	no.	%[a]	no.	%[a]	no.	%[a]	no.	%[a]
Contacts from the Time of								
Studies	54	33.1	17	20.7	5	13.9	9	15.2
Doct./Post Doc. at Univ.	37	22.7	17	20.7	6	16.7	21	35.6
Professional Activities in								
Research Establishments	11	6.7	6	7.3	3	8.3	21	35.6
Business	61	37.4	42	25.8	22	61.1	8	13.6
Total	163	100	82	100	36	100	59	100

Note: a percentage figure as column %

Table B.21 Selected characteristics of sample research establishments (1996)

	All		Research Intensive[a]		Non-Research Intensive[b]	
	no.	%	no.	%	no.	%
Status						
University	94	84.7	49	84.5	28	90.3
Public Non-Univ. Organisations[c]	16	14.4	9	15.5	3	9.7
Total	110	100	58	100	31	100
Employment Size						
≤ 19	66	76.7	42	75.0	24	80.0
20 – 49	9	10.5	6	10.7	3	10.0
≥ 50	11	12.8	8	12.8	3	10.0
Total	86	100	56	100	30	100

Notes: a more than 50% of the total time budget of all scientific staff devoted to basic or applied research

b less than 50% of the total time budget of all scientific staff devoted to basic or applied research

c including business-related research centres

Table B.22 Reasons for collaboration with research establishments (1996)

	Research Establishments in Science Fields											
	Architect. Construct., Surveying		Biology, Chemistry Medicine		Mathem., Inform., Physics		Electrotechn. Mechanical Engineering		Economics, Social and Geosciences		Total	
	no.	%[a]	no.	%[a]	no.	%[a]	no.	%[a]	no.	%[a]	no.	%[a]
Insufficient Own Equipment	10	71	2	18	7	29	1	25	7	70	27	43
Insufficient Own Personnel Capacity	5	36	3	27	10	42	3	75	5	50	26	41
Financial Sponsorship only Avail. for Collab.	11	79	7	63	15	63	2	50	5	50	40	63
Ideas for Research Work/Thematic Additions	2	14	6	55	13	54	2	50	3	30	26	41
Raising Own Profile	7	50	4	36	9	38	2	50	5	50	27	43
n	14		11		24		4		10		63	

Note: a percentage figure as % of cases

Table B.23 Branch structure of business clients (1996)

	Research Establishments in Science Fields											
	Architect.; Construct., Surveying		Biology, Chemistry Medicine		Mathem., Inform., Physics		Electrotechn. Mechanical Engineering		Economics, Social and Geosciences		Total	
	no.	%[a]	no.	%[a]	no.	%[a]	no.	%[a]	no.	%[a]	no.	%[a]
Energy and Mining	2	22	2	17	5	21	2	50	1	17	12	22
Basic Metals and Metal Products	2	22	4	33	3	13	2	50	0	0	11	20
Chemicals	0	0	8	67	2	8	0	0	0	0	10	18
Electrical & Optical Equipment	1	11	0	0	9	36	4	10	0	0	14	26
Computers	0	0	0	0	6	25	2	50	0	0	8	15
Plastics & Rubber	1	11	1	8	0	0	1	25	0	0	3	5
Machinery & Transport	2	22	1	8	4	17	3	75	0	0	10	18
Wood, Paper & Printing	0	0	0	0	1	4	0	0	1	17	2	4
Textiles & Clothing	0	0	0	0	2	8	0	0	0	0	2	4
Food Industry	0	0	4	33	4	17	0	0	0	0	8	15
Service Industry	5	56	1	8	12	50	1	25	6	100	25	45
n	9		12		24		4		6		55	

Note: a percentage figure as % of cases

Table B.24 Primary reasons for collaboration with businesses (1996)

	Research Establishments in Science Fields											
	Architect.; Construct., Surveying		Biology, Chemistry Medicine		Mathem., Inform., Physics		Electrotechn., Mechanical Engineering		Economics, Social and Geosciences		Total	
	no.	%[a]	no.	%[a]	no.	%[a]	no.	%[a]	no.	%[a]	no.	%[a]
Enables Costly Research Projects	9	69	9	75	9	37	3	75	2	33	32	54
Receipt of Funds Tied to Cooperation	10	77	8	67	12	50	3	75	4	67	37	63
Reduces Dependency on Public Contracts	8	61	7	58	16	67	3	75	3	33	37	63
Practice-Oriented Impulses for the Research Project	11	84	8	67	20	83	3	75	4	67	46	78
Creation of Jobs for Scientific Qualification	2	15	3	25	5	21	1	25	2	33	13	22
Use of Businesses' Capacities	7	54	4	33	10	42	3	75	2	33	26	44
n	13		12		24		4		6		59	

Note: a percentage figure as % of cases

Table B.25 Primary channels for establishing contacts (1996)

	Research Establishment in Science Fields											
	Architect.; Construct., Surveying		Biology, Chemistry Medicine		Mathem., Inform., Physics		Electrotech., Mechanical Engineering		Economics, Social and Geosciences		Total	
	no.	%[a]	no.	%[a]	no.	%[a]	no.	%[a]	no.	%[a]	no.	%[a]
Through												
Congresses/Fairs	2	15.4	3	25.0	5	20.8	2	50.0	1	16.7	13	22.0
Specialist Magazines	0	0.0	0	0.0	2	8.3	0	0.0	2	33.3	4	6.8
Data Banks	0	0.0	1	8.3	1	4.2	0	0.0	1	16.7	3	5.1
Associations	2	15.4	2	16.7	3	12.5	1	25.0	1	16.7	9	15.3
Transfer Establ./ Contact Agencies	0	0.0	1	8.3	1	4.2	0	0.0	1	16.7	3	5.1
Approach by Businesses	9	69.2	7	58.3	15	62.5	4	100	5	83.3	40	67.8
Personal Contact of Staff	12	92.3	11	91.7	20	83.3	4	100	6	100	53	89.8
n	13		12		24		4		6		59	

Note: a percentage figure as % of cases

Table B.26 Problems during collaboration with businesses (1996)

	Manufacturing Firms		Producer Service Firms		Total	
	no.	%[a]	no.	%[a]	no.	%[a]
Problems with Project Leadership	7	18.4	1	16.7	8	18.2
Budgeted Cost Overrun	13	34.2	1	16.7	14	31.2
Coordination is Difficult	19	50.0	3	50.0	22	50.0
Different Capabilities	23	60.5	4	66.7	27	61.4
Lack of Schedule Effectiveness	19	50.0	2	33.3	21	47.7
n	38		6		44	

Note: a percentage figure as % of cases

Table B.27 Barriers before collaborating with businesses/enterprises (1996)

	Cooperation with					
	Manufacturing Firms		Producer Service Firm		Total	
	no.	%[a]	no.	%[a]	no.	%[a]
Lacking of Actual Contact Partner	17	41.5	1	14.3	18	37.5
Financial Budget of Businesses Insufficient	29	70.7	6	85.7	35	72.9
Fear of Knowledge Drain to:						
the Businesses	17	41.5	2	28.6	19	39.6
the Research Establishments	10	24.4	1	14.3	11	22.9
n	103		6		109	

Note: a percentage figure as % of cases

Table B.28 Relevance of personal contacts (1996)

	Metropolitan Scale		National Scale		International Scale	
	no.	%[a]	no.	%[a]	no.	%[a]
Contacts from the Time of						
Studies	37	43.5	17	20.0	7	8.2
Ph.D./Post Doc Studies	33	38.8	28	32.9	32	37.6
Professional Activities in						
Research Institutions	44	51.8	33	38.8	38	44.7
Business	24	28.2	19	22.3	10	11.8

Note: a percentage figure as % of cases

APPENDIX C

The Stockholm Metropolitan Innovation System

Manufacturing Sector:	Tables C.1 – C.12
Producer Service Sector:	Tables C.13 – C.20
Science and Research Sector:	Tables C.21 – C.28

Table C.1 Selected characteristics of manufacturing sample firms (1996)

	Sample Firms with					
	Regular R&D[b]		No R&D[c]		Total	
	no.	%	no.	%	no.	%
Corporate Status						
Single Establishment	41	38.7	107	59.8	148	21.9
Multi-Establishment	65	61.3	72	40.2	137	48.1
Main Plant	*56*	*52.8*	*41*	*22.9*	*97*	*34.0*
Branch Plant	*9*	*8.5*	*31*	*17.3*	*40*	*14.1*
Total	106	100	179	100	285	100
Employment Size						
≤ 49	79	50.6	214	78.1	293	68.1
50 – 99	28	17.9	29	10.6	57	13.3
100 – 499	36	23.1	28	10.2	64	14.9
≥ 500	13	8.3	3	1.1	16	3.7
Total	156	100	274	100	430	100
R&D Expenditure [a]						
None	5	3.8	69	35.8	74	22.8
0.1 – 3.4	40	30.3	71	36.8	111	34.2
3.5 – 7.9	44	33.3	30	15.5	74	22.8
≥ 8	43	32.6	23	11.9	66	20.3
Total	132	100	193	100	325	100

Notes: a annual average of 1994-1996
b regular R&D firms are permanently engaged in research or development process
c no R&D firms are only occasionally or never engaged in research or development process

Table C.2 R&D of manufacturing firms (1996) disaggregated by size classes

	Firm Size by Employees									
	≤ 49		50 – 99		100 – 499		≥ 500		Total	
	no.	%[b]	no.	%[b]	no.	%[b]	no.	%[b]	no.	%[b]
***R&D Expenses* (in % of turnover)**										
No R&D	59	27.8	9	19.1	6	11.1	1	8.3	75	23.1
0.1 - 3.4	68	32.1	12	25.5	28	51.9	3	25.0	111	34.2
3.5 - 7.9	41	19.3	15	31.9	12	22.2	5	41.7	73	22.5
≥ 8	44	20.8	11	23.4	8	14.8	3	25.0	66	20.3
Total	212	100	47	100	54	100	12	100	325	100

Notes: a average 1994-1996
b percentage figure as column %

Table C.3 Innovativeness of manufacturing firms (1996) by age

	Year of Founding /Establishment									
	Before 1969		1970-1979		1980-1989		Since 1990		All	
	no.	%[a]	no.	%[a]	no.	%[a]	no.	%[a]	no.	%[a]
Firms										
Non-Innovative	64	33.3	30	15.6	37	19.3	61	31.8	192	100
Innovative[b]	89	35.9	37	14.9	57	23.0	65	26.2	248	100
Total	153	34.8	67	15.2	94	21.4	126	28.6	440	100

Notes: a percentage figure as row %
b firms with product innovations in 1994-1996

Table C.4 Manufacturing firms with product innovations (1994 - 1996)

	Corporate Status							
	Single Establishment		Main Plant		Branch Plant		Total	
	no.	%[a]	no.	%[a]	no.	%[a]	no.	%[a]
None	72	63.2	22	19.3	20	17.5	114	100
Further Development of Products[b]	29	43.3	27	40.3	11	16.4	67	100
New Products[c]	25	52.1	18	37.5	5	10.4	48	100
Total	126	55.0	67	29.3	36	15.7	229	100

Notes: a percentage figure as row %
b more than 50% of the products were substantially developed further during the years 1994 to 1996
c more than 50% of the were newly introduced in the product programme 1994 to 1996

Table C.5 Sources of external information for product and process innovation (1996)

	Product Innovation Activities		Process Innovation Activities	
	no.	%[a]	no.	%[b]
Sources				
Customers	231	93.5	58	33.0
Suppliers	135	54.7	134	76.1
Competitors	166	67.2	71	40.3
Research Institutions	64	25.9	55	31.3
Producer Services	46	18.6	53	30.1
Fairs/Exhibitions	172	69.5	112	63.6
Specialised Literature	146	59.1	119	67.6
Media	62	25.1	33	18.8
Internet	43	17.4	19	10.8
n	247		176	

Notes: a percentage of all firms with product innovation activities
b percentage of all firms with process innovation activities

Table C.6 Motives for exercising network activities (1996)

	Network Activities with					
	Research Institutions		Producer Service Providers		Other Manufacturing Firms[b]	
	no.	%[a]	no.	%[a]	no.	%[a]
Motives						
Risks/Cost Reduction	11	7.4	28	15.7	91	37.3
New Technological Opportunities	84	56.8	59	23.2	157	64.3
Know-How-Takeover	72	48.7	104	58.4	123	50.4
Financial Resources	10	6.8	48	27.0	17	7.0
Funding Requirements	42	28.4	42	23.6	19	7.8
n	148		178		244	

Notes: a percentage of manufacturing firms cooperating with the corresponding network partner
b relation of manufacturing firms with manufacturing suppliers, customers or competitors

Table C.7 Problems with exercising network activities (1996)

	Network Activities with					
	Research Institutions		Producer Service Providers		Other Manufacturing Firms[b]	
	no.	%[a]	no.	%[a]	no.	%[a]
Problems						
Problem with Project Management	19	24.7	50	29.6	73	33.6
Budgeted Cost Overrun	15	19.5	95	56.3	62	28.6
Unintended Knowledge Drain	18	23.4	34	20.1	21	9.7
Coordination Difficult	23	29.9	56	33.2	66	30.4
Different Capability	16	20.8	87	51.4	94	43.3
Confidential Relation/ Secrecy	25	32.5	29	17.2	68	31.3
Loss of Independence	7	9.1	15	8.9	15	6.9
Lack of Schedule Effectiveness	9	11.7	30	17.8	141	65.0
n	77		169		217	

Notes: a percentage of manufacturing firms cooperating with the corresponding network partner
b relation of manufacturing firms with manufacturing suppliers, customers or competitors

Table C.8 Network activities of manufacturing firms with customers (1996)

	Metro-politan Region		Rest of Mälardalen		National Scale		European Union		Mondial Scale		Total	
	no.	% [b]	no.	% [b]	no.	% [b]	no.	% [b]	no.	% [b]	no.	% [b]
Pre-competitive Stage												
Information Exchange												
Strong Ties [a]	99	25 (40)	74	19 (38)	112	28 (40)	69	17 (40)	44	11 (40)	398	100 (40)
Weak Ties [a]	18	24 (18)	21	28 (23)	20	27 (17)	11	15 (16)	5	7 (13)	75	100 (18)
Identification of New Ideas												
Strong Ties [a]	87	26 (35)	69	20 (36)	92	27 (33)	54	16 (31)	35	10 (32)	337	100 (34)
Weak Ties [a]	30	22 (30)	26	19 (28)	40	29 (34)	26	19 (38)	14	10 (37)	136	100 (33)
R&D												
Strong Ties [a]	64	24 (26)	50	19 (26)	73	27 (26)	49	18 (28)	30	11 (28)	266	100 (27)
Weak Ties [a]	53	26 (52)	45	22 (49)	59	29 (50)	31	15 (46)	19	9 (50)	207	100 (50)
Competitive Stage												
Prototype Development												
Strong Ties [a]	79	25 (38)	61	19 (41)	95	30 (41)	54	17 (37)	33	10 (36)	322	100 (39)
Weak Ties [a]	38	25 (26)	34	23 (25)	37	25 (23)	26	17 (28)	16	11 (29)	151	100 (25)
Pilot Projects												
Strong Ties [a]	65	26 (32)	44	17 (30)	70	28 (30)	45	18 (31)	30	12 (33)	254	100 (31)
Weak Ties [a]	52	24 (36)	51	23 (37)	62	28 (38)	35	16 (37)	19	9 (34)	219	100 (37)
Marketing												
Strong Ties [a]	62	25 (30)	43	17 (29)	69	28 (29)	47	19 (32)	28	11 (31)	249	100 (30)
Weak Ties [a]	55	25 (38)	52	23 (38)	63	28 (39)	33	15 (35)	21	9 (38)	224	100 (38)

Notes: a number of firms with corresponding network activities
b percentage figure as row %, percentage figure as column % of respective innovation stage in parentheses

Table C.9 Network activities of manufacturing firms with manufacturing suppliers (1996)

	Metro-politan Region		Rest of Mälardalen		National Scale		European Union		Mondial Scale		Total	
	no.	%[b]	no.	%[b]	no.	%[b]	no.	%[b]	no.	%[b]	no.	%[b]
Pre-competitive Stage												
Information Exchange												
Strong Ties[a]	32	21 (43)	24	15 (41)	45	29 (38)	35	23 (37)	19	12 (41)	155	100 (39)
Weak Ties[a]	8	16 (18)	6	12 (19)	18	35 (26)	12	24 (26)	7	14 (22)	51	100 (23)
Identification of New Ideas												
Strong Ties[a]	26	19 (35)	21	15 (36)	43	31 (36)	33	24 (35)	15	11 (33)	138	100 (35)
Weak Ties[a]	14	21 (31)	9	13 (29)	20	29 (29)	14	21 (30)	11	16 (34)	68	100 (30)
R&D												
Strong Ties[a]	17	17 (23)	14	14 (24)	32	31 (27)	27	26 (28)	12	12 (26)	102	100 (26)
Weak Ties[a]	23	22 (51)	16	15 (52)	31	30 (45)	20	19 (43)	14	13 (44)	104	100 (47)
Competitive Stage												
Prototype Development												
Strong Ties[a]	25	20 (51)	19	15 (53)	36	29 (46)	29	23 (43)	17	13 (61)	126	100 (49)
Weak Ties[a]	15	19 (21)	11	14 (20)	27	34 (24)	18	23 (24)	9	11 (18)	80	100 (22)
Pilot Projects												
Strong Ties[a]	18	21 (37)	12	14 (33)	27	31 (35)	23	26 (34)	7	8 (25)	87	100 (34)
Weak Ties[a]	22	18 (31)	18	15 (33)	36	30 (32)	24	20 (32)	19	16 (38)	119	100 (33)
Marketing												
Strong Ties[a]	6	13 (12)	5	11 (14)	15	33 (19)	15	33 (22)	4	9 (14)	45	100 (17)
Weak Ties[a]	34	21 (48)	25	16 (46)	48	30 (43)	32	20 (43)	22	14 (44)	161	100 (45)

Notes: a number of firms with corresponding network activities
b percentage figure as row %, percentage figure as column % of respective innovation stage in parentheses

Table C.10 Network activities of manufacturing firms with producer service providers (1996)

	Metro-politan Region no.	%[b]	Rest of Mälardalen no.	%[b]	National Scale no.	%[b]	European Union no.	%[b]	Mondial Scale no.	%[b]	Total no.	%[b]
Pre-competitive Stage												
Information Exchange												
Strong Ties[a]	37	36	22	22	28	27	6	6	9	9	102	100
		(32)		(32)		(30)		(32)		(33)		(31)
Weak Ties[a]	42	42	21	21	25	25	7	7	4	4	99	100
		(35)		(35)		(37)		(35)		(33)		(35)
Identification of New Ideas												
Strong Ties[a]	46	38	22	18	35	29	8	7	10	8	121	100
		(39)		(32)		(38)		(42)		(37)		(37)
Weak Ties[a]	33	41	21	26	18	23	5	6	3	4	80	100
		(28)		(35)		(27)		(25)		(25)		(29)
R&D												
Strong Ties[a]	34	34	25	25	29	29	5	5	8	8	101	100
		(29)		(36)		(32)		(26)		(30)		(31)
Weak Ties[a]	45	45	18	18	24	24	8	8	5	5	100	100
		(38)		(30)		(36)		(40)		(42)		(36)
Competitive Stage												
Prototype Development												
Strong Ties[a]	46	42	18	17	25	23	10	9	10	9	109	100
		(52)		(50)		(48)		(53)		(43)		(50)
Weak Ties[a]	33	36	25	27	28	30	3	3	3	3	92	100
		(22)		(27)		(26)		(15)		(19)		(24)
Pilot Projects												
Strong Ties[a]	22	37	10	17	14	24	5	8	8	14	59	100
		(25)		(28)		(27)		(26)		(35)		(27)
Weak Ties[a]	57	40	33	23	39	27	8	6	5	4	142	100
		(39)		(35)		(36)		(40)		(31)		(37)
Marketing												
Strong Ties[a]	21	41	8	16	13	25	4	8	5	10	51	100
		(24)		(22)		(25)		(21)		(22)		(23)
Weak Ties[a]	58	39	35	23	40	27	9	6	8	5	150	100
		(39)		(38)		(37)		(45)		(50)		(39)

Notes: a number of firms with corresponding network activities
b percentage figure as row %, percentage figure as column % of respective innovation stage in parentheses

Table C.11 Network activities of manufacturing firms with research institutions (1996)

	Metro-politan Region		Rest of Mälardalen		National Scale		European Union		Mondial Scale		Total	
	no.	%[b]	no.	%[b]	no.	%[b]	no.	%[b]	no.	%[b]	no.	%[b]
Pre-competitive Stage												
Information Exchange												
Strong Ties[a]	43	40	11	10	33	31	15	14	5	5	107	100
		(39)		(41)		(36)		(37)		(38)		(38)
Weak Ties[a]	12	32	8	21	15	39	2	5	1	3	38	100
		(22)		(27)		(29)		(20)		(20)		(25)
Identification of New Ideas												
Strong Ties[a]	34	39	9	10	28	32	12	14	4	5	87	100
		(31)		(33)		(30)		(29)		(31)		(31)
Weak Ties[a]	21	36	10	17	20	34	5	9	2	3	58	100
		(39)		(33)		(38)		(50)		(40)		(38)
R&D												
Strong Ties[a]	34	38	7	8	31	34	14	16	4	4	90	100
		(31)		(26)		(34)		(34)		(31)		(32)
Weak Ties[a]	21	38	12	22	17	31	3	5	2	4	55	100
		(39)		(40)		(33)		(30)		(40)		(36)
Competitive Stage												
Prototype Development												
Strong Ties[a]	18	35	5	10	17	33	9	18	2	4	51	100
		(56)		(56)		(50)		(50)		(40)		(52)
Weak Ties[a]	37	39	14	15	31	33	8	9	4	4	94	100
		(28)		(29)		(28)		(24)		(31)		(28)
Pilot Projects												
Strong Ties[a]	12	32	3	8	14	38	6	16	2	5	37	100
		(38)		(33)		(41)		(33)		(40)		(38)
Weak Ties[a]	43	40	16	15	34	31	11	10	4	4	108	100
		(32)		(33)		(31)		(33)		(31)		(32)
Marketing												
Strong Ties[a]	2	20	1	10	3	30	3	30	1	10	10	100
		(6)		(11)		(9)		(17)		(20)		(10)
Weak Ties[a]	53	39	18	13	45	33	14	10	5	4	135	100
		(40)		(38)		(41)		(42)		(38)		(40)

Notes: a number of firms with corresponding network activities
b percentage figure as row %, percentage figure as column % of respective innovation stage in parentheses

Table C.12 Network activities of manufacturing firms with competitors in form of producer networks (1996)

	Metro-politan Region		Rest of Mälardalen		National Scale		European Union		Mondial Scale		Total	
	no.	% [b]	no.	% [b]	no.	% [b]	no.	% [b]	no.	% [b]	no.	% [b]
Pre-competitive Stage												
Information Exchange												
Strong Ties [a]	13	29 (32)	9	20 (32)	16	36 (34)	5	11 (29)	2	4 (18)	45	100 (31)
Weak Ties [a]	8	27 (36)	7	23 (35)	1	3 (4)	6	20 (38)	8	27 (42)	30	100 (30)
Identification of New Ideas												
Strong Ties [a]	13	27 (32)	10	21 (36)	15	31 (32)	6	13 (35)	4	8 (36)	48	100 (33)
Weak Ties [a]	8	22 (36)	6	16 (30)	12	32 (50)	5	14 (31)	6	16 (32)	37	100 (37)
R&D												
Strong Ties [a]	15	29 (37)	9	18 (32)	16	31 (34)	6	12 (35)	5	10 (45)	51	100 (35)
Weak Ties [a]	6	18 (27)	7	21 (35)	11	32 (46)	5	15 (31)	5	15 (26)	34	100 (34)
Competitive Stage												
Prototype Development												
Strong Ties [a]	11	22 (44)	9	18 (53)	16	32 (41)	9	18 (41)	5	10 (38)	50	100 (43)
Weak Ties [a]	10	29 (26)	7	20 (23)	11	31 (26)	2	6 (18)	5	14 (29)	35	100 (25)
Pilot Projects												
Strong Ties [a]	9	25 (36)	5	14 (29)	12	33 (31)	6	17 (27)	4	11 (31)	36	100 (31)
Weak Ties [a]	12	24 (32)	11	22 (35)	15	31 (36)	5	10 (45)	6	12 (35)	49	100 (35)
Marketing												
Strong Ties [a]	5	17 (20)	3	10 (18)	11	37 (28)	7	23 (32)	4	13 (31)	30	100 (26)
Weak Ties [a]	16	29 (42)	13	24 (42)	16	29 (38)	4	7 (36)	6	11 (35)	55	100 (40)

Notes: a number of firms with corresponding network activities
b percentage figure as row %, percentage figure as column % of respective innovation stage in parentheses

Table C.13 Selected characteristics of sample producer service firms (1996)

	Sample Firms									
	All		Innov. Firm[a]		Non Innov. Firm[b]		Technical Orientated[c]		Business Orientated[d]	
	no.	%	no.	%	no.	%	no.	%	no.	%
Corporate Status										
Single Establishment	122	57.0	86	53.1	36	69.2	61	50.4	61	65.6
Multi-Establishment	92	43.0	76	46.9	16	30.8	60	49.6	32	34.4
Main Plant	*57*	*26.6*	*50*	*30.9*	*7*	*13.5*	*36*	*29.8*	*21*	*22.6*
Branch Plant[e]	*35*	*16.3*	*26*	*16.1*	*9*	*17.3*	*24*	*19.9*	*11*	*11.9*
Total	214	100	162	100	52	100	121	100	93	100
Employment Size										
≤ 19	128	54.5	87	49.2	41	70.7	64	48.5	64	62.1
20 – 49	62	26.4	50	28.2	12	20.7	33	25.0	29	28.2
50 – 249	34	14.5	30	16.9	4	6.9	25	18.9	9	8.7
≥ 250	11	4.7	10	5.6	1	1.7	10	7.6	1	1.0
Total	235	100	177	100	58	100	132	100	103	100

Notes: a introducing new services and/or organisational innovations
b spending more than 20% of turnover in R&D activities
c spending less than 20% of turnover in R&D activities
d belonging to the sectors of computer software and technical consultancy
e belonging to the sectors of business consultancy and market research/advertising

Table C.14 R&D expenses of producer service firms (1996) disaggregated by size classes

		Firm Size by Employees									
		≤ 19		20 – 49		50 – 249		≥ 250		Total	
		no.	%[b]	no.	%[b]	no.	%[b]	no.	%[b]	no.	%[b]
R&D Expenses											
(in % of turnover)[a]											
None		3	3.9	0	0.0	0	0.0	0	0.0	3	2.0
	c	100		0.0		0.0		0.0		100	
0.1 – 3.4		49	64.5	35	81.4	20	80.0	8	88.5	112	73.2
	c	43.8		31.3		17.9		7.1		100	
3.5 – 7.9		13	17.1	7	16.3	4	16.0	1	11.1	25	16.3
	c	52.0		28.0		16.0		4.0		100	
≥ 8		11	14.5	1	2.3	1	4.0	0	0.0	13	8.5
	c	84.6		7.7		7.7		0.0		100	
Total		76	100	43	100	25	100	9	100	153	100
	c	49.7		28.1		16.3		5.5		100	

Notes: **a average 1994 to 1996**
b percentage figure as column %
c percentage figure as row %

Table C.15 Innovativeness of producer service firms (1996) by age

	Year of Founding / Establishment									
	Before 1970		1970-1979		1980-1989		Since 1990		All	
	no.	%[a]	no.	%[a]	no.	%[a]	no.	%[a]	no.	%[a]
Firms										
Non-Innovative	11	18.0	3	4.9	20	32.8	27	44.3	61	100
Innovative[b]	25	14.3	19	10.9	61	34.9	70	40.0	175	100
Total	36	15.3	22	9.3	81	34.3	97	41.1	236	100

Notes: **a percentage figure as row %**
b firms with product innovations in 1994-1996

Table C.16 Firms with service innovations (1994 - 1996)

	Single Establishment		Main Plant		Branch Plant		Total	
	no.	%[a]	no.	%[a]	no.	%[a]	no.	%[a]
None	36	69.2	7	13.5	9	17.3	52	100
New Services	51	50.5	36	35.6	14	13.9	101	100
Substantially Improved Services	52	49.5	38	36.2	15	14.3	105	100
New or Substantially Improved Methods of Services Provision	46	54.1	25	29.4	14	16.5	85	100

Note: a percentage figure as row %

Table C.17 Types of innovation support for manufacturing firms (1996) by producer service firms

	Computer Software		Technical Consultancy		Business Consultancy		Market Research and Advertising		Total	
In %	yes	no	yes	no	yes	no	yes	no	yes	no
Product Innovation	41.6	58.4	65.7	34.3	40.0	60.0	28.6	71.4	42.1	57.9
Process Innovation	60.4	39.6	45.7	54.3	56.4	43.6	22.4	77.6	49.6	50.4
Organisat. Innovations	50.5	49.5	40.0	60.0	61.8	38.2	30.6	69.4	47.5	52.5
New Sales Markets with Exist. Products	36.6	63.4	31.4	68.6	38.2	61.8	42.9	57.1	37.5	62.5

Table C.18 Important factors for successful collaboration with manufacturing clients (1996)

	Computer Software		Technical Consultancy		Business Consultancy		Market Research and Advertising		Total	
	no.	%[a]	no.	%[a]	no.	%[a]	no.	%[a]	no.	%[a]
Factors										
Spatial Proximity	47	18.2	20	21.1	19	16.7	14	18.4	100	18.4
b		47.0		20.0		19.0		14.0		100
Frequent Personal Contacts	73	28.3	28	29.5	37	32.5	23	30.3	161	26.7
b		45.4		17.4		23.0		14.3		100
Existence of Similar Qualifications	66	25.6	22	23.8	31	27.1	19	25.0	138	25.4
b		47.8		15.9		22.5		13.8		100
Good Knowledge of Client´s Industry	72	27.9	25	26.4	27	23.7	20	26.3	144	26.5
b		50.0		17.4		18.8		13.9		100
Total	258	100	95	100	114	100	76	100	543	100
b		47.5		17.5		21.0		14.0		100

Notes: a percentage figure as column %
b percentage figure as row %

Table C.19 Establishments of contacts (1996)

	Computer Software		Technical Consultancy		Business Consultancy		Market Research and Advertising		Total	
	no.	%[a]	no.	%[a]	no.	%[a]	no.	%[a]	no.	%[a]
Contacts through										
Advertising/ Marketing	30	11.6	10	12.7	19	14.8	30	25.9	89	15.3
Conference/ Meetings	25	9.7	8	10.1	14	10.9	6	51.7	53	9.1
Fairs	13	50.4	3	3.8	2	1.6	0	0.0	18	3.1
References/ Recommendations	73	28.3	22	27.8	32	25.0	31	26.7	158	27.2
Contact Agencies	6	2.3	2	2.5	5	3.9	5	4.3	18	3.1
Existing Contacts w. Former Colleagues	76	29.5	28	35.4	43	33.6	32	27.6	179	30.8
Existing Contacts w. Former Employees	35	13.6	6	7.6	13	10.1	12	10.3	66	11.4
Total	258	145.4	79	100	128	100	116	146	581	100

Note: a percentage in % of responses

Table C.20 Relevance of personal contacts (1996)

	Metropolitan Scale		National Scale		International Scale		Not Important	
	no.	%[a]	no.	%[a]	no.	%[a]	no.	%[a]
Contacts from the Time of								
Studies	39	26.9	49	31.0	9	21.4	70	21.3
Doct./Post Doc. at Univ.	23	15.9	31	19.6	5	11.9	101	30.8
Professional Activities in								
Research Establishments	9	6.2	19	12.0	3	7.1	129	39.3
Business	74	51.0	59	37.3	25	59.3	28	8.5
Total	145	100	158	100	42	100	328	100

Note: a percentage figure as column %

Table C.21 Selected characteristics of sample research establishments (1996)

	All		Research Intensive[a]		Non-Research Intensive[b]	
	no.	%	no.	%	no.	%
Status						
University	69	76.7	52	73.2	16	88.9
Public Non-Univ. Organisations[c]	21	23.3	19	26.8	2	11.1
Total	90	100	71	100	16	100
Employment Size						
≤ 19	84	51.9	66	50.8	18	56.3
20 – 49	45	27.8	40	30.8	5	15.6
≥ 50	33	20.4	24	18.5	9	28.1
Total	162	100	130	100	32	100

Notes: a more than 50% of the total time budget of all scientific staff devoted to basic or applied research
b less than 50% of the total time budget of all scientific staff devoted to basic or applied research
c including business-related research centres

Table C.22 Reasons for collaboration with research establishments (1996)

	Research Establishments in Science Fields											
	Architect. Construct., Surveying		Biology, Chemistry Medicine		Mathem., Inform., Physics		Electrotechn. Mechanical Engineering		Economics, Social and Geosciences		Total	
	no.	%[a]	no.	%[a]	no.	%[a]	no.	%[a]	no.	%[a]	no.	%[a]
Insufficient Own Equipment	4	16	31	18	6	13	5	11	9	13	55	16
Insufficient Own Personnel Capacity	4	16	30	18	6	13	6	14	12	17	58	16
Financial Sponsorship only Avail. for Collab.	4	16	37	22	11	24	9	20	15	21	76	21
Ideas for Research Work/Thematic Additions	8	32	39	23	11	24	14	32	19	27	91	26
Raising Own Profile	5	20	33	19	11	24	10	23	15	21	74	21
n	25		170		45		44		70		354	

Note: a percentage figure as % of cases

Table C.23 Branch structure of business clients (1996)

	Research Establishments in Science Fields											
	Architect.; Construct., Surveying		Biology, Chemistry Medicine		Mathem., Inform., Physics		Electrotechn. Mechanical Engineering		Economics, Social and Geosciences		Total	
	no.	%[a]	no.	%[a]	no.	%[a]	no.	%[a]	no.	%[a]	no.	%[a]
Energy and Mining	1	5	3	4	2	11	2	11	1	2	9	5
Basic Metals and Metal Products	1	5	3	4	1	6	1	5	1	2	7	4
Chemicals	1	5	13	18	7	39	1	5	10	21	32	18
Electrical & Optical Equipment	3	16	6	8	1	6	4	21	6	13	20	11
Computers	4	21	7	10	1	6	3	16	10	21	25	14
Plastics & Rubber	0	0	3	4	0	0	0	0	2	42	5	3
Machinery & Transport	3	16	4	6	2	11	0	0	4	8	13	7
Wood, Paper & Printing	1	5	6	8	1	6	4	21	4	8	16	9
Textiles & Clothing	0	0	1	1	0	0	0	0	0	0	1	1
Food Industry	0	0	9	13	0	0	0	0	3	6	12	7
Service Industry	5	26	16	23	3	17	4	21	7	15	35	20
n	19		71		18		19		48		175	

Note: a percentage figure as % of cases

Table C.24 Primary reasons for collaboration with businesses (1996)

	Research Establishments in Science Fields											
	Architect.; Construct., Surveying		Biology, Chemistry Medicine		Mathem., Inform., Physics		Electrotechn., Mechanical Engineering		Economics, Social and Geosciences		Total	
	no.	%[a]	no.	%[a]	no.	%[a]	no.	%[a]	no.	%[a]	no.	%[a]
Enables Costly Research Projects	3	12	37	23	7	16	8	13	3	12	58	18
Receipt of Funds Tied to Cooperation	6	24	33	20	9	21	14	23	6	24	68	21
Reduces Dependency on Public Contracts	6	24	26	16	10	24	9	14	7	28	58	18
Practice-Oriented Impulses for the Research Project	2	8	33	20	8	19	14	23	2	8	49	15
Creation of Jobs for Scientific Qualification	3	12	12	7	4	9	10	16	5	20	34	11
Use of Businesses' Capacities	5	20	21	13	5	12	7	11	2	8	40	13
n	25		162		43		62		25		317	

Note: a percentage figure as % of cases

Table C.25 Primary channels for establishing contacts (1996)

	Research Establishment in Science Fields										Total	
	Architect.; Construct., Surveying		Biology, Chemistry Medicine		Mathem., Inform., Physics		Electrotech., Mechanical Engineering		Economics, Social and Geosciences			
	no.	%[a]	no.	%[a]	no.	%[a]	no.	%[a]	no.	%[a]	no.	%[a]
Through												
Congresses/Fairs	2	25.0	16	32.7	5	31.3	4	21.1	4	33.3	31	29.8
Specialist Magazines	1	12.5	17	34.7	3	18.8	3	15.8	3	25.0	27	26.0
Data Banks	0	0.0	1	2.0	1	6.3	0	0.0	0	0.0	2	1.9
Associations	2	25.0	13	26.5	3	18.8	4	21.1	2	16.7	24	23.1
Transfer Establ./ Contact Agencies	1	12.5	2	4.1	2	12.5	4	21.1	0	0.0	9	8.7
Approach by Businesses	6	75.0	28	57.1	8	50.0	10	52.6	6	50.0	58	55.8
Personal Contact of Staff	7	87.5	38	77.6	13	81.3	19	100	10	83.3	87	83.7
n	18		48		22		29		40		157	

Note: a percentage figure as % of cases

Table C.26 Problems during collaboration with businesses (1996)

	Manufacturing Firms		Producer Service Firms		Total	
	no.	%[a]	no.	%[a]	no.	%[a]
Problems with Project Leadership	3	10.7	0	0.0	3	10.3
Budgeted Cost Overrun	5	17.9	0	0.0	5	17.2
Coordination is Difficult	9	32.1	1	100	10	34.5
Different Capabilities	23	82.1	1	100	24	82.8
Lack of Schedule Effectiveness	7	25.0	1	100	8	27.6
n	28		1		29	

Note: a percentage figure as % of cases

Table C.27 Barriers before collaborating with businesses/enterprises (1996)

	Cooperation with					
	Manufacturing Firms		Producer Service Firm		Total	
	no.	%[a]	no.	%[a]	no.	%[a]
Lacking of Actual Contact Partner	21	55.3	2	50.0	23	54.8
Financial Budget of Businesses Insufficient	18	47.4	3	75.0	21	50.0
Fear of Knowledge Drain to:						
the Businesses	8	21.1	0	0.0	8	19.0
the Research Establishments	6	15.8	0	0.0	6	14.3
n	38		4		42	

Note: a percentage figure as % of cases

Table C.28 Relevance of personal contacts (1996)

	Metropolitan Scale		National Scale		International Scale	
	no.	%[a]	no.	%[a]	no.	%[a]
Contacts from the Time of						
Studies	37	21.1	40	20.5	14	11.2
Ph.D./Post Doc Studies	49	28.0	58	29.7	43	34.4
Professional Activities in						
Research Institutions	62	35.4	68	34.9	55	44
Business	27	15.4	29	14.9	13	10.4
Total	175	100	195	100	125	100

Note: a percentage figure as % of cases

References

Aiginger, K. and Peneder, M. (1997): Qualität und Defizite des Industriestandorts Österreich. Austrian Institute of Economic Research, Vienna

Alderman, N. (1999): Local Product Development Trajectories: Engineering Establishments in Three Contrasting Regions. In: Malecki, E.J., Oinas, P. (eds.): *Making Connections. Technological Learning and Regional Economic Change*, pp. 79-107. Ashgate, Aldershot.

Anderstig, C. and Hårsman, B. (1986): On Occupation Structure and Location Pattern in the Stockholm Region, *Regional Science and Urban Economics* 16, 25-35

Andersson, Å. (1985): Creativity and Regional Development, *Papers of the Regional Science Association* 56, 5-20

Andersson, Å.E. and Andersson, D. (eds.) (2000): *Gateways to the Global Economy*. Edward Elgar, Cheltenham.

Anselin, L., Varga, A. and Acs, Z. (1997): Local Geographic Spillovers between University Research and High Technology Innovations, *Journal of Urban Economics* 42, 422-448

Archibugi, D. (1996): National Innovation Systems. A Comparative Analysis, *Research Policy* 25, 838-842

Asheim, B.T. and Cooke, P. (1999): Local Learning and Interactive Innovation Networks in a Global Economy. In: Malecki, E.J., Oinas, P. (eds.): *Making Connections. Technological Learning and Regional Economic Change*, pp. 145-178. Ashgate, Aldershot.

Audretsch, D.B. (1999): Is the German Economic Model Still Viable? *Wirtschaftspolitische Blätter* 46(3), 276-285

Axelsson, S. (1992): Firms and Regions in Networks – The Role of the Stockholm Region in the Swedish Economic Production System. PhD Dissertation, Royal Institute of Technology, Stockholm

Bacaria, J. and Borras, S. (1998): The Catalan Innovation System: Governing Rapid Changes. In: Braczyk, H.-J., Cooke, P., Heidenreich, M. (eds.): *Regional Innovation Systems. The Role of Governances in a Globalized World*, pp. 72-98. UCL Press, London

Bailly, A. S. and Coffey, W. J. (1991): Flexible Production and Producer Services, Paper Presented at the Conference on *Regional and Urban Restructuring in Europe*, RURE, Lisbon, 17-19 February 1991

Becker, W., Peters, J. (1997): Hochschulen als Faktor für Wirtschaft und Arbeitsmarkt einer Region. In: Ermert, K. (ed.): *Hochschule und Region.*

Wirkungen und Wechselwirkungen, pp. 47-59 [=Loccumer Protokolle 17/97], Evangelische Akademie Loccum, Rehburg-Loccum

Bienaymé, A. (1986): The Dynamics of Innovation, *International Journal of Technology Management* 1, 133-159

Bienefeld, B.J. (1995): *Wettbewerbsfähigkeit und Internationalisierungseffekte am Beispiel Katalanischer Industrieunternehmungen*. Vandenhoeck & Ruprecht, Göttingen

Braczyk, H.-J., Cooke, P. and Heidenreich, M. (eds.) (1998): *Regional Innovation Systems. The Role of Governances in a Globalized World*. UCL Press, London

Braun, E. and Macdonald, S. (1982): *Revolution in Miniature, the History and Impact of Semiconductor Electronics*. Cambridge University Press, Cambridge [MA]

Breschi, S. and Malerba, F. (1997): Sectoral Innovation Systems: Technological Regimes, Schumpeterian Dynamics, and Spatial Boundaries. In: Edquist, C. (ed.): *Systems of Innovation. Technologies, Institutions and Organizations*, pp. 130-156. Pinter, London

Britton, S. (1990): The Role of Services in Production, *Progress in Human Geography* 14, 529-546

Brooks, H. (1994): The Relationship between Science and Technology, *Research Policy* 23, 477-486

Caixa de Catalunya (1995): Anuari Economic Comarcal 1995. Barcelona

Camagni, R. (ed.) (1991): *Innovation Networks: Spatial Perspectives*. Pinter, London

Capello, R. (2000): *Urban Innovation and Collective Learning: Theory and Evidence from Five Metropolitan Cities in Europe*, Paper Presented at the International Workshop on *Knowledge, Complexity and Urban Innovation Systems*, Vienna, 1-3 July 2000

Caracostas, P. and Soete, L. (1997): The Building of Cross-Border Institutions in Europe: Towards a European System of Innovation? In: Edquist, C. (ed.): *Systems of Innovation. Technologies, Institutions and Organizations*, pp. 395-419. Pinter, London

Caravaca, B.I. and Sanchez Lechuga, P. (1995): Cambios Socioeconomicos, Desempleo y Desequilibrios Territoriales en España, *Estudios Regionales* 42, 15-52

Carlsson, B. (ed.) (1995): *Technological Systems and Economic Performance: The Case of Factory Automation*. Kluwer, Dordrecht, Boston

Castells, M. (1996): *The Rise of the Network Society*. Blackwell, Oxford

CEP (1992): Anuari del Serveis a les Empreses de Catalunya. Barcelona

Charles, D. and Goddard, J. (1997): Higher Education and Employment – Linking Universities with their Regional Industrial Base, Paper Prepared for the Thematic Seminar on *Territorial Employment Pacts*, Osterlund [Sweden]

Chesnais, F. (1988): Technical Co-operation Agreements Between Firms. STI Review 4, Organisation for Economic Co-operation and Development, Paris

CIDEM (1996): Informacion General. Barcelona

Cohen, W. M. and Levinthal, D. A. (1989): Innovation and Learning: The Two Faces of R&D, *Economic Journal* 99, 569-596

Cooke, P. (1998): Introduction: Origins of the Concept. In: Braczyk, H.-J., Cooke, P., Heidenreich, M. (eds.): *Regional Innovation Systems. The Role of Governances in a Globalized World*, pp. 2-25. UCL Press, London

Cooke, P. (1996): Reinventing the Region: Firms, Clusters and Networks in Economic Development. In: Daniels, P.W., Lever, W.F. (eds.): *The Global Economy in Transition*, pp. 310-327. Longman, Harlow, Essex

Cooke, P., Boekholt, P. and Tödtling, F. (2000): *The Governance of Innovations in Europe. Regional Perspectives on Global Competitiveness*. Pinter, London

Cooke, P., Gomez, M. and Etxebarria, G. (1997): Regional Innovation Systems: Institutional and Organisational Dimensions, *Research Policy* 26, 475-491

Crewe, L. (1996): Material Culture: Embedded Firms, Organizational Networks and the Local Economic Development of a Fashion Quarter, *Regional Studies* 30(3), 257-272

De Bresson, C. and Amesse, F. (1991): Networks of Innovators: A Review and Introduction to the Issue, *Research Policy* 20, 363-379

Dicken, P. (1998): *Global Shift. Transforming the World Economy*. The Guilford Press, New York, London

Dosi, G. (1988): Sources, Procedures, and Microeconomic Effects of Innovation, *Journal of Economic Literature* 26, 139-144

Dunning, J.H. (1997): *Governments, Globalisation and International Business*. Oxford University Press, Oxford, New York

Edquist, C. (1997a): *Systems of Innovation. Technologies, Institutions and Organizations*. Pinter, London

Edquist, C. (1997b): Systems of Innovation Approaches - Their Emergence and Characteristics. In: Edquist, C. (ed.): *Systems of Innovation. Technologies, Institutions and Organizations*, pp. 1-35. Pinter, London

Edquist, C. and Johnson, B. (1997): Institutions and Organizations in Systems of Innovation. In: Edquist, C. (ed.): *Systems of Innovation. Technologies, Institutions and Organizations*, pp. 41-63. Pinter, London

Edquist, C. and Rees, G. (2000): Learning Regions and Cities: Learning in Regional Innovation Systems – A Conceptual Framework, Paper Presented at the International Workshop on *Knowledge, Complexity and Innovation Systems*, Vienna, 1-3 July 2000

Escorsa, P. (1994): La Politica de Apoyo a la Innovacion: El Caso Catalunya, Grup S.A., Barcelona

Escorsa, P. and Valls, J. (1992): La Recerca i la Tecnologia. Quaderns de Competivitat 12, Grup S.A., Barcelona

Europäische Kommission (1999): Sechster Periodischer Bericht über die Sozio-ökonomische Entwicklung der Regionen der Europäischen Union. Luxembourg

Evangelista, R., Sandven, T., Sirilli, G. and Smith, K. (1997): *Innovation Expenditures in European Industry*. EIMS Publication no. 48. Luxembourg

Feldman, M. (1994): *The Geography of Innovation*. Kluwer, Dordrecht, Boston

Fischer, M.M. (2001): Innovation, Knowledge Creation and Systems of Innovation, *The Annals of Regional Science* 35 [forthcoming]

Fischer, M.M. (1999): The Innovation Process and Network Activities of Manufacturing Firms. In: Fischer, M.M., Suarez-Villa, L., Steiner, M. (eds.):

Innovation, Networks and Localities, pp. 11-27. Springer, Berlin, Heidelberg, New York

Fischer, M.M. (1990): The Economic Role of Producer Services, WSG Discussion Paper 5/90, Department of Economic Geography & Geoinformatics, Vienna University of Economics and Business Administration

Fischer, M.M. and Fröhlich, J. (eds.) (2001): *Knowledge, Complexity and Innovation Systems*. Springer, Berlin, Heidelberg, New York

Fischer, M.M. and Nijkamp, P. (1985): Developments in Exploratory Discrete Spatial Data and Choice Analysis, *Progress in Human Geography* 9, 515-551

Fischer, M.M. and Varga, A. (2000): Technological Innovation and Interfirm Co-operation. An Exploratory Analysis Using Survey Data from Manufacturing Firms in the Metropolitan Region of Vienna, *The International Journal of Technology Management* [in press]

Fischer, M.M., Suarez-Villa, L. and Steiner, M. (eds.) (1999): *Innovation, Networks and Localities*. Springer, Berlin, Heidelberg, New York

Florax, R. (1992): *The University: A Regional Booster: Economic Impacts of Academic Knowledge Infrastructure*. Avebury, Aldershot

Freeman, C. (1991): Networks of Innovators: A Synthesis of Research Issues, *Research Policy* 20, 499-514

Freeman, C. (1987): *Technology and Economic Performance: Lessons from Japan*. Pinter, London

Fritsch, M. and Schwirten, C. (1998): Öffentliche Forschungseinrichtungen im Regionalen Innovationssystem, *Raumforschung und Raumordnung* 4, 253-263

Fritsch, M., Koschatzky, K., Schätzl, L. and Sternberg, R. (1998): Regionale Innovationspotentiale und Innovative Netzwerke, *Raumforschung und Raumordnung* 4, 243-252

Garcia, E. (1989): El Reto Empresarial Español - la Empresa Española y su Competividad. Madrid

Generalitat de Catalunya (1999): Anuario Estadistico 1999. Barcelona

Generalitat de Catalunya (1997): Anuario Estadistico 1997. Barcelona

Generalitat de Catalunya (1995): Anuario Estadistico 1995. Barcelona

Generalitat de Catalunya (1992): La Recerca i la Tecnologia. Barcelona

Granovetter, M. (1985): Economic Action and Social Structure: The Problem of Embeddedness, *American Journal of Sociology* 93(3), 481-510

Gregersen, B. and Johnson, B. (1997): Learning Economies, Innovation Systems and European Integration, *Regional Studies* 31(5), 479-490

Guinet, J. (1997): Knowledge Flows in National Innovation Systems. In: OECD (ed.): *Industrial Competitiveness in the Knowledge-Based Economy. The New Role of Governments*, pp. 173-178. Organisation for Economic Co-operation and Development, Paris

Hall, P. (1997): The Political Economy of Adjustment in Germany. In: Naschold, F., Soskice, D., Hancké, B., Jürgens, U. (eds): *Ökonomische Leistungsfähigkeit und Institutionelle Innovation*, pp. 293-317. Edition Sigma, Berlin

Hall, P. (1987): The Geography of High Technology: An Anglo-American Comparison. In: Brotchie, J., Hall, P., Newton, P. (eds.): *The Spatial Impact of Technological Change*, pp. 141-156. Croom Helm, London, New York, Sydney

Hall, P. (1986): The Theory and Practice of Innovation Policy: An Overview. In: Hall, P. (ed.): *Technology Innovation and Economic Policy*, pp. 1-34. St. Martin's Press, New York

Hall, P. and Markusen, A. (eds.) (1985): *Silicon Landscapes*. Allen & Unwin, Boston

Hall, P., Prud'homme, R. and Snickars, F. (1994): *The Impacts of the Dennis Agreement: Regional Development*. The Swedish National Rail Administration, The Stockholm County Council, The Greater Stockholm Rapid Transit Authority, The City of Stockholm and The Swedish National Road Administration

Håkansson, H. (1987): *Industrial Technological Development: A Network Approach*. Croom Helm, London, New York, Sydney

Harrison, B. (1982): Rationalization, Restructuring and Industrial Organization in Older Regions. The Economic transformation of New England since World War II, Working Paper 72, Joint Center for Urban Studies, MIT, Harvard

Hatzichronoglou, T. (1997): Revision of the High-Technology Sector and Product Classification, STI Working Papers 1997/2. Organisation for Economic Co-operation and Development, Paris

Hohlfeld, P. (1995): Integration ungleich entwickelter Wirtschaftsräume – das Beispiel des EU-Beitritts Spaniens, *RWI-Mitteilungen* 3, 237-255

Hudson, R. (1999): The Learning Economy, the Learning Firm and the Learning Region: A Sympathetic Critique of the Limits to Learning, *European Urban and Regional Studies* 6(1), 59-72

Hutschenreiter, G., Knoll, N., Paier, M. and Ohler, F. (1998): Austrian Report on Technology 1997. Study by the Austrian Institute for Economic Research (WIFO) and the Austrian Research Centre Seibersdorf (ARCS). Austrian Institute for Economic Research, Vienna

Instituto Nacional de Estadistica (1995): Encuesta Industrial. Madrid

ISI (1993): Technologieprofil der Region Rhein–Main. Gutachten für den Umlandverband Frankfurt, Karlsruhe

Jaffe, A. (1989): Real Effects of Academic Research, *American Economic Review* 79, 957-970

Johannisson, B. (1991): University Training for Entrepreneurship: Swedish Approaches, *Entrepreneurship and Regional Development* 3, 67-82

Johnson, B. (1997): Systems of Innovation: Overview and Basic Concepts. In: Edquist, C. (ed.): *Systems of Innovation. Technologies, Institutions and Organizations*, pp. 36-40. Pinter, London

Jörg, L. (1997): Comparative Analysis of the Scientific Performance of Austria: The Costs of Universalism. OEFZS-4809, Austrian Research Centre Seibersdorf

Kamann, D.J. and Nijkamp, P. (1990): Technogenesis: Incubation and Diffusion. In: Cappellin, R., Nijkamp, P. (eds.): *The Spatial Context of Technological Development*, pp. 257-302. Avebury, Aldershot

Kanter, R. (1995): *World Class: Thriving Locally in the Global Economy*. Simon and Schuster Publishers, New York

Kirat, T. and Lung, Y. (1999): Innovations and Proximity. Territories as Loci of Collective Learning Processes, *European Urban and Regional Studies* 6(1), 27-38

Kline, S. J. and Rosenberg, N. (1986): An Overview of Innovation. In: Landau, R., Rosenberg, N. (eds.): *The Positive Sum Strategy*, pp. 275-305. National Academy Press, Washington,

Koschatzky, K. (1995): *Regionale Innovations- und Technologieförderung*. Physica Verlag, Karlsruhe

Kuhn, S. (1982): *Computer Manufacturing in New England – Structure, Location and Labour in a Growing Industry*. Joint Center for Urban Studies, MIT, Cambridge [MA]

Kuntze, O.E. (1990): *Vorbereitung zehn Westeuropäischer Industrieländer auf den EG-Binnenmarkt*. IFO Studien zur Europäischen Wirtschaft 1, München

Lassnigg, L. and Pollan W. (1996): Das Österreichische Qualifizierungssystem im Internationalen Vergleich, *WIFO-Monatsberichte* 69, 763-780

Lundvall, B.-Å. (1994): The Global Unemployment Problem and National Systems of Innovation. In: O'Doherty, D. (ed.): *Globalisation, Networking and Small Firm Innovation*, pp. 35-48. Graham & Trotman, London

Lundvall, B.-Å. (ed.) (1992): *National Systems of Innovation: Towards a Theory of Innovation and Interactive Learning*. Pinter, London

Lundvall, B.-Å. (1988): Innovation as an Interactive Process: From User-Producer Interaction to the National System of Innovations. In: Dosi, G., Freeman, C., Nelson, R., Silverberg, G., Soete, L. (eds.): *Technical Change and Economic Theory*, pp. 349-369. Pinter, London

Malecki, E.J. (1997): *Technology & Economic Development*. Longman, Essex (2nd edition)

Malecki, E.J. (1991): *Technology and Economic Development: The Dynamics of Local, Regional, and National Change*. Longman, London

Malecki, E.J. and Oinas, P. (eds.) (1999): *Making Connections. Technological Learning and Regional Economic Change*. Ashgate, Aldershot

Malecki, E.J. and Veldhoen, M.E. (1993): Network Activities, Information and Competitiveness in Small Firms, *Geografiska Annaler* 75B, 131-147

Mansfield, E. (1968): *Industrial Research and Technological Change*. W.W. Norton, New York

Mansfield, E. and Lee, J.-Y. (1996): The Modern University: Contributor to Industrial Innovation and Recipient of Industrial R&D Support, *Research Policy* 25, 1047-1058

Mansfield, E., Romeo, A., Schwartz, M., Teece, D., Wagner, S. and Brach, P. (1982): *Technology Transfer, Productivity, and Economic Policy*. W.N. Norton, New York

Markusen, A., Hall, P. and Glasmeier, A. (1986): *High Tech America, The What, How, Where and Why of the Sunrise Industries*. Allen & Unwin, London

Martinelli, F. (1991): A Demand-Oriented Approach to Understanding Producer Services. In: Daniels, P. W., Moulaert, F. (eds.): *The Changing Geography of Advanced Producer Services*, pp. 154-185. Belhaven Press, London, New York

Maskell, P. and Malmberg, A. (1999): The Competitiveness of Firms and Regions. 'Ubiquitification' and the Importance of Localized Learning, *European Urban and Regional Studies* 6(1), 9-25

Massey, D., Quintas, P. and Wield, D. (1992): *High-Tech Fantasies: Science Parks in Society, Science and Space*. Routledge, London

Metcalfe, J.S. (1992): Variety, Structure and Change: An Evolutionary Perspective on the Competitive Process, *Revue d'Économie Industrielle* 59(1), 46-61

Meyer-Krahmer, F. (1985): Innovative Behaviour and Regional Indigenous Potential, *Regional Studies* 19, 523-534

Molero, J. and Buesa, M. (1996a): Innovatory Activity in Spanish Firms: Regular versus Occasional Patterns, Paper Presented at the International Conference on *Management and Technologies*, Madrid, 12-14 June 1996

Molero, J. and Buesa, M. (1996b): Technological Strategies of MNCs in Intermediate Countries: The Case of Spain, Papers of the Complutense University. Madrid

Morgan, K. (1997): The Learning Region: Institutions, Innovation and Regional Renewal, *Regional Studies* 31(5), 491-503

Myers, M.B. and Rosenbloom, R.S. (1996): Rethinking the Role of Industrial Research. In: Rosenbloom, R.S., Spencer, W.J. (eds.): *Engines of Innovation: US Industrial Research at the End of an Era*, pp. 209-228. Harvard Business School Press, Cambridge [MA]

Nelson, R.R. (ed.) (1993): *National Innovation Systems. A Comparative Analysis*. Oxford University Press, New York, Oxford

Nelson, R. and Winter, S. (1982): *An Evolutionary Theory of Economic Change*. Harvard University Press, Cambridge [MA]

Nijkamp, P. (1990): *The Urban Incubator Hypothesis*. Working Paper, Department of Spatial Economics, Free University, Amsterdam

Nijkamp, P., Oirschot, G. van and Oosterman, A. (1994): Knowledge Networks, Science Parks and Regional Development: An International Comparative Analysis of Critical Success Factors. In: Cuadrado-Roura, J.R., Nijkamp, P., Salva, P. (eds.): *Moving Frontiers: Economic Restructuring, Regional Development and Emerging Networks*, pp. 225-246. Ashgate, Aldershot

Noisi, J., Saviotti, P., Bellon, B. and Crow, M. (1993): National Systems of Innovation. In Search of a Workable Concept, *Technology in Society* 15(2), 207-227

Nonaka, I. and Takeuchi, H. (1995): *The Knowledge-Creating Company. How Japanese Companies Create the Dynamics of Innovation*. Oxford University Press, Oxford, New York

NUTEK (1997): *The National Swedish Innovation System – A Quantitative Study*. Nutek Report 12 (1997), Stockholm

OECD (1999): *Benchmarking Knowledge Based Economies*. Organisation for Economic Co-operation and Development, Paris

OECD (1995a): *National Systems for Innovation Financing*. Organisation for Economic Co-operation and Development, Paris

OECD (1995b): *OECD Economic Surveys 1994-1995: Austria*. Organisation for Economic Co-operation and Development, Paris

OECD (1994): *National Systems of Innovation: General Conceptual Framework*. DSTI/STP/TIP 94(4). Organisation for Economic Co-operation and Development, Paris

OECD (1992): *Technology and Economy: The Key Relationships*. Organisation for Economic Co-operation and Development, Paris

OECD (1988): *Reviews of National Science and Technology Policy: Austria*. Organisation for Economic Co-operation and Development, Paris

Ohmae, K. (1995): *The End of the Nation State*. Free Press, New York

Oinas, P. and Malecki, E.J. (1999): Spatial Innovation Systems. In: Malecki, E.J., Oinas, P. (eds.): *Making Connections. Technological Learning and Regional Economic Change*, pp. 7-33. Ashgate, Aldershot

Parker D. and Zilberman, D. (1993) University Technology Transfers: Impacts on Local and U. S. Economies, *Contemporary Policy Issues* 11, 87-99

Pavitt, K. (1984): Sectoral Patterns of Technical Change: Towards a Taxonomy, *Research Policy* 13, 343-373

Polanyi, M. (1996): *The Tacit Dimension*. Routledge & Kegan Paul, London

Porter, M.E. and Fuller, M.B. (1986): Coalitions and Global Strategy. In: Porter, M.E. (ed.): *Competition in Global Industries*, pp. 315-343. Harvard Business School Press, Boston

Powell, W.W. (1990): Neither Market nor Hierarchy: Network Forms of Organization. In: Staw, B.M., Cummings, L.L. (eds.): *Research in Organizational Behavior*, pp. 295-335. JAI Press, Greenwich CT

Rogers, E. and Larsen, J. (1984): *Silicon Valley Fever: Growth of High-Technology Culture*. Basic Books, New York

Romer, P. (1990): Endogenous Technical Progress, *Journal of Political Economy* 98,71-103

Sabel, C. (1994): Bootstrapping Reform: Rebuilding Firms, the Welfare State and Unions, Paper Presented to the Conféderation des Syndicats Nationaux, Montreal

Saviotti, P.P. (1998): On the Dynamics of Appropriability of Tacit and of Codified Knowledge, *Research Policy* 26, 843-856

Saviotti, P.P. (1988): Information, Entropy and Variety in Technoeconomic Development, *Research Policy* 17, 89-103

Saxenian, A. (1994): *Regional Advantage: Culture and Competition in Silicon Valley and Route 128*. Harvard University Press, Cambridge [MA]

Schibany, A. (1998): Co-operative Behaviour of Innovative Firms in Austria. Austrian Institute for Economic Research, Vienna

Scott, A.J. (1988): *New Industrial Spaces*. Pion, London

Sirilli, G. and Evangelista, R. (1998): Technological Innovation in Services and Manufacturing: Results from Italian Surveys, *Research Policy* 27, 881-899

Snickars, F. (2000): The Role of the Randstad Region in the European Urban System. In: Andersson, Å. E., Andersson, D. (eds.): *Gateways to the Global Economy*, pp. 296-310. Edward Elgar, Cheltenham

Sörlin, S. and Törnqvist, G. (2000): *Knowledge for Welfare – the Universities and the Transformation of Sweden*. SNS Publishing Company, Stockholm

Storper, M. (1997): *The Regional World. Territorial Development in a Global World.* The Guilford Press, New York, London

Strambach, S. (1999): Innovation Processes and the Role of Knowledge-Intensive Business Services, Paper Presented at the ISI-Conference on *Innovation Networks – Concepts and Challenges in the European Perspective*, Karlsruhe, 18-19 November 1999

Suarez-Villa, L. (2000): *Innovation and the Rise of Technocapitalism.* Rowman & Littlefield, Lanham, Boulder New York, Oxford

Suarez-Villa, L. (1989): *The Evolution of Regional Economies. Entrepreneurship and Macroeconomic Change.* Praeger, New York

Teece, D.J. (1986): Profiting from Technological Innovation: Implications for Integration, Collaboration, Licensing and Public Policy, *Research Policy* 15, 285-305

Teece, D. (1981): The Market for Know-how and the Efficient International Transfer of Technology, *Annals of the American Academy of Political and Social Science* 458, 81-96

Tijssen, R.J.W. (1998): Quantitative Assessment of Large Heterogeneous R&D Networks: The Case of Process Engineering in the Netherlands, *Research Policy* 26, 791-809

Traxler, J., Fischer, M.M., Nöst, A. and Schubert, U. (1991): Producer Service Networks in the Metropolitan Region of Vienna. In: Bergman, E., Maier, G., Tödtling, F. (eds.): *Regions Reconsidered*, pp. 197-212. Mansell, London

Wiig, H. and Wood, M. (1997): What Comprises a Regional Innovation System? Theoretical Base and Indicators. In: Simmie, J. (ed.): *Innovation, Networks and Learning Regions*, pp. 66-98. Jessica Kingsley, London

Varga, A. (1998): *University Research and Regional Innovation. A Spatial Econometric Analysis of Academic Technology Transfers.* Kluwer, Dordrecht, Boston

Zairi, M. (1992): Competitive Benchmarking: An Executive Guide. TQM Practitioner Series. Technical Communications (Publishing) Ltd., Letchworth

Zucker, L., Darby, M. and Brewer, M. (1998): Intellectual Human Capital and the Birth of U.S. Biotechnology Enterprises, *American Economic Review* 88, 290-306

List of Figures

List of Tables

Appendix A: The Vienna Metropolitan Innovation System

Appendix B: The Barcelona Metropolitan Innovation System

Appendix C: The Stockholm Metropolitan Innovation System

Subject Index

Author Index

MIX
Papier aus verantwortungsvollen Quellen
Paper from responsible sources
FSC® C105338

If you have any concerns about our products, you can contact us on
ProductSafety@springernature.com

In case Publisher is established outside the EU, the EU authorized representative is:
Springer Nature Customer Service Center GmbH
Europaplatz 3, 69115 Heidelberg, Germany

Printed by Libri Plureos GmbH
in Hamburg, Germany